The Decrease in Velocity of α-Particles in Passing through Matter.

By E. MARSDEN, M.Sc., Lecturer, and T. S. TAYLOR, Ph.D., John Harling Fellow, University of Manchester.

(Communicated by Prof. E. Rutherford, F.R.S. Received April 22,—Read May 1, 1913.)

RUTHERFORD BACK-SCATTERING AS A TOOL TO DETERMINE ELECTRONIC STOPPING POWERS IN SOLIDS*

R. BEHRISCH AND B. M. U. SCHERZER
Max-Planck-Institut für Plasmaphysik, EURATOM-Association, D-8046 G
(Received August 10, 1973)

Range distribution of implanted ions in SiO_2, Si_3N_4, a

W.K. Chu*, B.L. Crowder†, J.W. Mayer*, and J.F. Ziegler†
IBM Thomas J. Watson Research Center, Yorktown Heights, New York 10598
(Received 24 January 1973; in final form 12 March 1973)

DETERMINATION OF IMPLANTATION PROFILES IN SOLI
BY SECONDARY ION MASS SPECTROMETRY *

J. MAUL, F. SCHULZ and K. WITTMAACK
Gesellschaft für Strahlen- und Umweltforschung mbH,
...isch-Technische Abteilung D-8042 Neuherberg, Germany.
...ed 3 July 1972

IMPLANTATION OF BORON IN SILICON *)

BY

W. K. HOFKER

P. HVELPLUND

ENERGY LOSS AND STRAGGLING OF 100–500 keV ...WITH $2 \leq Z_1 \leq 12$ IN ...IOUS GASES

Range measurements in oriented tungsten single crystals. III. The influence of temperature on the maximum range[1]

J. A. DAVIES, L. ERIKSSON,[2] AND J. L. WHITTON[3]
Chalk River Nuclear Laboratories, Chalk River, Ontario
Received November 22, 1967

RECHERCHES EXPÉRIMENTALES SUR LE PASSAGE DES RAYONS α A TRAVERS LA MATIÈRE

...FOR LOW-ENERGY ... IN CARBON ... CROSS ... ATOMS WITH ... ...CKWORTH

J. H. ORMROD AND ...

The range of Rn^{222} ions of keV energies in aluminum and tungsten

By I. BERGSTRÖM, J. A. DAVIES, B. DOMEIJ and J. UHLE

...the Invalidity of Bragg's Rule in Stopping Cross ... Molecules for Swift Li Ions *

W. Neuwirth, W. Pietsch, K. Richter, and U. Hauser
Erstes Physikalisches Institut, Universität zu Köln
Received September 15, ...

EXPERIMENTAL INVESTIGATION OF HIGHER-ORDER Z_1 CORRECTIONS TO THE BETHE STOPPING-POWER FORMULA

H. H. ANDERSEN, J. F. BAK, H. KNUDSEN, P. MØLLER PETERSEN and B. R. NIELSEN
...titute of Physics, University of Aarhus, DK-8000 Aarhus C, Denmark
...October 1976

phys. stat. sol. 29, 403 (1968)
Subject classification: 11, 21.6
Physik-Department der Technischen Hochschule München
Section Physik der Universität München (b)

Energy Loss of ⟨110⟩ Channelled α-Recoil Atom

By
K. IBEL (a),[2] and R. SIZMANN (b)

...ION-IMPLANTED DONORS IN SILICON*

R. W. Bower, R. Baron, J. W. Mayer, and O. J. Marsh '75
Hughes Research Laboratories, Malibu, California
(Received 29 June 1966)

MEASUREMENT OF THE ENERGY LOSS OF GERMANIUM ATOMS TO ELECTRONS IN GERMANIUM AT ENERGIES BELOW 100 keV*

C. Chasman, K. W. Jones, and R. A. Ristinen
Brookhaven National Laboratory, ...
(Received 9 July 1...)

VOLUME 103, NUMBER 1

...He Stopping Cross Section of Solids for Protons, 50–600 keV*

M. BADER,† R. E. PIXLEY, F. S. MOZER, AND W. WHALING
Kellogg Radiation Laboratory, California Institute of Technology, Pasadena, California
(Received March 30, 1956)

A SECTIONING TECHNIQUE FOR COPPER, SILVER AND ITS APPLICATION TO PENETRATION AND STUDIES

T. ANDERSEN
Institute of Chemistry, University of Aarhus, Aarhus C, Denmark

AND

G. SORENSEN
... of Aarhus, Aarhus C, Denmark

Channeling Effects in the Energy Loss of 3–11-MeV Protons in Silicon and Germanium Single Crystals*

B. R. APPLETON
...e Laboratories, Murray Hill, New Jersey and Rutgers, The State University, New Brunswick, New Jers...

AND

C. ERGINSOY
Brookhaven National Laboratory, Upton, New York

AND

W. M. GIBSON
...ratories, Murray Hill, New Jersey and Rutgers, The State Universi...
(Received 31 March 1967)

..., SER. 2, VOL. 1, PRINTED IN GREAT BRITAIN

Čerenkov energy loss of muons in water

D. ALEXANDER, K. M. PATHAK† and M. G. TH...
Department of Physics, University of Durham
MS. received 8th May 1968

1 S

PHYSICAL REVIEW B

RESOLUTION OF THE Σ^--MASS ANOMALY

Walter H. Barkas, John N. Dyer,* and Harry H. Heckman
...Radiation Laboratory, University of California, Berkeley, California
(Received 29 May 1963)

Reprinted from THE PHYSICAL REVIEW, Vol. 149, No. 1, 244–245, 9 September 1966
Printed in U. S. A.

Alpha-Particle Stopping Cross Section of Solids from 0.3 to 2.0

W. K. Lin, H. G. Olson, and D. Powers
...aylor University, Waco, Texas 76703
... February 1973

Energieverluste und Ladungszustände schwerer Ionen beim Durchgang durch Materie

K. BETHGE, P. SANDNER und H. SCHMIDT

Mitteilungen aus dem Institut für Radium-forschung.

Some Heavy-Ion Stopping Powers*

C. D. MOAK AND M. D. BROWN†
Oak Ridge National Laboratory, Oak Ridge, Tennessee
(Received 11 April 1966)

Thin Solid Films, 19 (1973) 187–194
© Elsevier Sequoia S.A., Lausanne—Printed in Switzerland

187

VOLUME 106, NUMBER 5

LXIX.

die Reichweite der α-Strahlen in Flüssigkeiten

von

Wilhelm Michl.†

(Mit 18 Textfiguren.)

...rom THE PHYSICAL REVIEW, Vol. 153, ...am 9. Juli 1914.)
Printed in U. S. A.

TECHNIQUE FOR ACCURATELY MEASURING THE STO... CROSS SECTIONS OF REACTIVE METALS BY ION BACK-SCATTERING: He IONS IN ERBIUM*

R. A. LANGLEY AND R. S. BLEWER
...YSICAL REVIEW
Sandia Laboratories, Albuquerque, N. ...
(Received August 10, 1973)

Stopping Power of Some Metallic Elements for 10.8-Mev Protons

V. C. BURKIG AND K. R. MACKENZIE
Department of Physics, University of California, Los Angeles, California
(Received February 19, 1957)

Stopping Power of Be, Al, Cu, Ag, Pt, and Au for 5–12-MeV Protons and Deuterons

H. H. ANDERSEN, C. C. HANKE, H. SØRENSEN, AND P. VAJDA
Physics Department, Atomic Energy Commission, Research Establishment Risø, Roskilde, Denmark
(Received 1 August 1966)

Bibliography and Index of Experimental Range and Stopping Power Data

Other Titles in This Series

Bibliography and Index of Experimental Range and Stopping Power Data

H.H. ANDERSEN

IBM-Research
Yorktown Heights, New York
10598   USA

and

Institute of Physics
Aarhus University
Aarhus C, Denmark
(Permanent Address)

Volume 2
of
The Stopping and Ranges of Ions in Matter

Organized by: J. F. Ziegler

PERGAMON PRESS
New York / Toronto / Oxford / Sydney / Frankfurt / Paris

Pergamon Press Offices:

U.S.A.	Pergamon Press Inc., Maxwell House, Fairview Park, Elmsford, New York 10523, U.S.A.
U.K.	Pergamon Press Ltd., Headington Hill Hall, Oxford OX3, OBW, England
CANADA	Pergamon of Canada, Ltd., 75 The East Mall, Toronto, Ontario M8Z 5WR, Canada
AUSTRALIA	Pergamon Press (Aust) Pty. Ltd., 19a Boundary Street, Rushcutters Bay, N.S.W. 2011, Australia
FRANCE	Pergamon Press SARL, 24 rue des Ecoles, 75240 Paris, Cedex 05, France
WEST GERMANY	Pergamon Press GmbH, 6242 Kronberg/Taunus, Frankfurt-am-Main, West Germany

---

Copyright © 1977 Pergamon Press Inc.

Library of Congress Cataloging in Publication Data

Andersen, Hans Henrik.
    Bibliography and index of experimental range and stopping data.

    (The Stopping and ranges of ions in matter ; v. 2)
    1.   Stopping power (Nuclear physics) -- Indexes.
2. Effective range (Nuclear physics) -- Indexes. 3. Ions
-- Scattering -- Indexes. I. Title. II. Series.
Z7144.N8A5 [QC794.6.S8]   016.5397'21   77-22415
ISBN 0-08-021604-8

Printed in the United States of America

Contents

The experimental investigation of the energy loss of charged particles in matter goes back more than 70 years.[1] Since then, more than 900 papers containing energy loss and range information for charged particles other than electrons have been published. A complete review of this profusive amount of experimental data has never been attempted and even partial presentations, as for instance of proton and alpha particle stopping powers,[2] are not up-to-data. The theoretical situation is somewhat less confusing. The high energy loss of light ions is well described by the Bethe theory[3] as may be seen from Fano's review. [4] At the lower end of the energy spectrum the Thomas-Fermi model of the atom constitutes a useful framework for the treatment of range and energy loss phenomena.[5,6] The intermediate energy region is not covered by any comprehensive theoretical treatment and even within the covered regions, detailed comparisons show deviations ("Z-oscillations," "shell-effects") from the predicted behavior. The theoretical situation has been reviewed recently by Sigmund.[7]

In the absence of a really detailed theoretical picture and comprehensive compilations of experimental data, a large amount of tables of semiempirical nature have appeared. These tables are most useful for rapid estimates and a number of more recent ones are listed in Table 1. In most cases, it will turn out to be useful also to consult the original experimental literature. It is hoped that the present bibliography and index will help retrieving the relevant data.

In most cases it will be necessary to consult several sources simultaneously. Bichsel[8] noted that stopping powers measured by different groups tended not to agree within their stated errors. This situation has not improved since then. Sørensen and Andersen[9] found proton stopping powers 2-3% higher than Ishiwari et al.[10] although both authors stated errors of the order of 0.4%. Chu and Powers[11] found helium stopping powers in argon differing from those of Hanke and Bichsel[12] by as much as 5%, which is far outside the combined stated errors. Andersen et al[13] found proton stopping powers in calcium up to 5% below those of Gorodetzky et al[14] although the latter stated errors of 0.1%. At lower energies the situation may grow worse as seen in Fig. 1 which presents a complete graph of published proton stopping powers in gold below 20 MeV.[15]

Table 1

Some stopping power and range tables.

H. Bichsel and C. Tschalaer: A range energy table for heavy particles in silicon. Nuclear Data Tables *A3*, 393 (1967) (Range, dE/dx. p,d,t, ^{3}He, ^{4}He, ^{7}Li. 1 MeV and up).

J. F. Janni: Calculations of energy loss, range, pathlength, straggling, multiple scattering and the probability of inelastic nuclear collisions for 0.1 to 1000 MeV protons. AFWL-TR 65-150 (1966) (Range; dE/dx. Protons in many elements. 100 keV and up).

L. C. Northcliffe and R. F. Schilling: Range and stopping power of heavy ions. Nuclear Data Tables *A7*, 233 (1970) ($0.0125 \leq E/M \leq 12$ MeV/amu. All elements, 24 target materials).

J. F. Gibbons, N. S. Johnson and S. W. Mylroie: Projected range statistics, 2nd ed., Halsted Press (1975) (Range straggling, 32 projectiles 10-1000 keV. Mainly semiconducting targets).

J. F. Ziegler and W. K. Chu: Stopping cross sections and backscattering factors for ^{4}He ions in matter. Atomic Data and Nuclear Data Tables *13*, 463 (1974) (dE/dx. 0.4-4.0 MeV α in all elements).

H. H. Andersen and J. F. Ziegler: Hydrogen Stopping Powers and Ranges in All Elements. Pergamon (1977). (dE/dx 10 keV - 20 MeV hydrogen ions in all elements. Also graphical representation of all published experimental data).

Many of the above mentioned discrepancies could not be resolved even by careful analysis of the relevant publications. Hence, it was not found profitable to write a critical review of the compiled data, but merely to present an index and a bibliography and let the user choose for himself. To help this choice, target phases (gasous, fluid, solid, crystalline, amorphous) have been characterized in the index for data where they may be important. But the user should always be critically aware of the possibility of impurities, texture and, for thin layers, pinholes influencing the published data.

Content of Tables

The present compilation started as a supplement to a report by Brown and Jarmie[16] indexing 87 references containing original experimental information. Hence, their layout has to a large extent been used here also. As was the case for Brown and Jarmie, the present work specifically excludes electron range and energy loss data. Fission fragments are included as projectiles, but no systematic search of the literature has been made. On the target side, Brown and Jarmie's coverage has been supplemented with nuclear emulsions. A literature search has been conducted, but as the topic is rather far away from the authors usual occupation, the coverage is most probably less complete than for other target materials. Biological target materials are not included.

Table V is the main bibliography. Literature is cited giving all authors, full title and references. For Russian literature, which has appeared in English translation, both the original and the translation reference is given. The title is given in the original language with the following exceptions:

1) Articles which have also appeared in a cover-to-cover English translation. Here the title of the translation is given.

2) Articles which have not been directly available to the present author, and where only a translated title is known. The original language will be stated in a parenthesis.

3) Articles where the original language may not be written in Latin characters. The original language will be stated in a parenthesis.

Finally, the table summarizes the data presented in the reference. S means stopping power and R range. Straggling is indicated by δS and δR. Range straggling will often indicate higher order moments or complete distributions also. Transverse - range straggling is denoted by $\delta R_{\perp}$. Measurements of partition of energy between atomic motion and electronic excitation will be indicated by $\eta(\varepsilon)$.[17] Projectiles and targets are, as in all tables, given by the usual chemical symbols. For hydrogen the notation p,d,t and H,D,T will also be used for ^1H, ^2H and ^3H. Similarly, ^4He will often be given as α. Information on specific isotopes and on target structure will be given if considered of importance and if retrievable from the publication.

Table III is an index of the content of the data listing in Table V. The data appear sorted with respect to target materials. Compound targets appear under the heading of all important components with some qualifications specified below. The following compounds are listed separately:

Air	No references under nitrogen and oxygen.
Emulsion	No references under specific components.
Minerals	No references under specific components. The group mainly contains data for mica.
Organic Materials	No reference under H_2, C and O_2. If the materials contain other elements, e.g., nitrogen and halides, specific reference will also be given for these elements.
Oxides	No references under oxygen, but reference for other components.
Water	No reference under H_2 and O_2.

In Table IV the material has been sorted according to projectiles. For each projectile the data appear in random order, mainly as acquired for the tabulations. This will hopefully not give difficulties except for the most abundant projectiles, viz. p and α. As most range measurements also contain information on range straggling, this information was considered redundant and has not be included in Tables III and IV.

Finally, Table VI is an author index which is necessary to retrieve the work of specific persons or groups.

The indices are probably most easily used by first referring to Figs. 2 to 7. Here the material has been sorted into three energy groups, E < 100 keV, 100 keV $\leq$ E < 3 MeV and 3 MeV $\leq$ E. This sorting corresponds roughly to accelerator types; the low energy region corresponds to work performed at classical isotope separators, the intermediate region to single-ended van de Graaff accelerators and the high energy region to a number of larger machines including tandems, cyclotrons and linacs. For each energy region, and for range and stopping power separately, a target vs projectile matrix is given. A dot indicates that at least one datum exists for the specific combination and that further search may be worthwhile. A given target entry indicates that measurements were either performed on the pure element or a compound target containing the element.

New Publications

Publications on energy loss and range keep appearing at a rate of approximately three per month. A bibliography like the present one will thus be outdated quite rapidly and rather useless if not supplemented. Hence, it appears to be important to indicate where new information is liable to appear. To this end, Table II states the number of articles that have appeared in the 17 most frequently used sources in the 1970's. Some extrapolation is probably possible and the table may serve as a clue to where new material may appear. The table represents 85 percent of the last seven years publications. The heavy use of conference proceedings is noted. They are nearly all from conferences either on ion implantation or atomic collisions in solids. The trend is that an increasing fraction of the publications appear in sources, most publishing in the fields of solid state physics and materials science.

Table II

Sources of literature on ranges and energy loss data in the period 1970-1976.

Conference Proceedings	63
Nucl. Instr. Methods	55
Phys. Rev.	47
Rad. Effects	46
Appl. Phys. Letters	19
J. Appl. Phys	16
Phys. Letters	14
Thin Solid Films	13
Phys. Stat. Sol.	11
Z. Physik	11
Appl. Physics	8
Can. J. Phys.	7
J. Electrochem. Soc.	6
Bull. Am. Phys. Soc.	6
Sov. Phys. Solid State	5
Jap. J. Appl. Phys	5
Health Physics	5

Acknowledgments

The author is most grateful to his management at IBM Research for allowing him to use a substantial part of his visit at the Thomas J. Watson Laboratory to finish this project which was started while he was with the Danish AEC and continued at the Institute of Physics, Aarhus University. The author also wants to thank Mrs. Eva Pedersen, Librarian, Danish AEC and Mrs. Inger Vibeke Nielsen, Librarian, Institute of Physics, Aarhus for extensive help in retrieving literature for the project over the last ten years. In the final stages several colleagues responded to a call for supplementary material and hence contributed to the completeness of the bibliography. Finally Linda Callahan, Barbara Juliano, and Linda Rubin must be thanked for their careful and patient typing of this lengthy manuscript.

References

1. W. H. Bragg and R. Kleeman, Phil Mag *10*, 318 (1905).

2. W. Whaling, Handbuch der Physik *34/2*, 193 (1958).

3. H. A. Bethe, Ann. Physik *5*, 325 (1930).

4. U. Fano, Ann. Rev. Nucl. Sci. *13*, 1 (1963).

5. J. Lindhard, M. Scharff and H. E. Schiøtt, Kgl. Danske Videnskab. Selskab Mat. Fys. Medd. *33*, No. 14 (1963).

6. J. Lindhard and M. Scharff, Phys. Rev. *124*, 128 (1961).

7. P. Sigmund, p. 1 in J. H. S. Dupuy (ed), *Radiation Damage Processes in Materials,* Noordhoff. Leyden, The Netherlands (1975).

8. H. Bichsel, p. 17 in U. Fano (ed), *Studies in Penetration of Charged Particles in Matter,* National Academy of Sciences, Washington D.C. (1964).

9. H. Sørensen and H. H. Andersen, Phys. Rev. B*8*, 1854 (1973).

10. R. Ishiwari, N. Shiomi, S. Shirai, T. Ohata and Y. Uemura, Bull. Inst. Chem. Res., Kyoto Univ. *49*, 390 (1971).

11. W. K. Chu and D. Powers, Phys. Rev. B*4*, 10 (1971).

12. C. C. Hanke and H. Bichsel, Kgl. Danske Videnskab. Selskab, Mat. Fys, Medd. *38*, No. 3 (1970).

13. H. H. Andersen, C. C. Hanke, H. Simonsen, H. Sørensen and P. Vajda, Phys. Rev. *175*, 389 (1968).

14. S. Gorodetzky, A. Chevallier, A. Pape, J. C. Sens, A. M. Bergdolt, M. Bres, and R. Armbruster, Nucl. Phys. A*91*, 133 (1967).

15. H. H. Andersen and J. F. Ziegler, *Hydrogen Stopping Powers and Ranges in All Elements*. Pergamon, (1977).

16. R. E. Brown and N. Jarmie, Los Alamos Scientific Laboratory Report No. LA-2156 (1958).

17. J. Lindhard, V. Nielsen, M. Scharff and P. V. Thomsen, Kgl. Danske Videnskab. Selskab. Mat. Fys. Medd. *33*, No. 10 (1963).

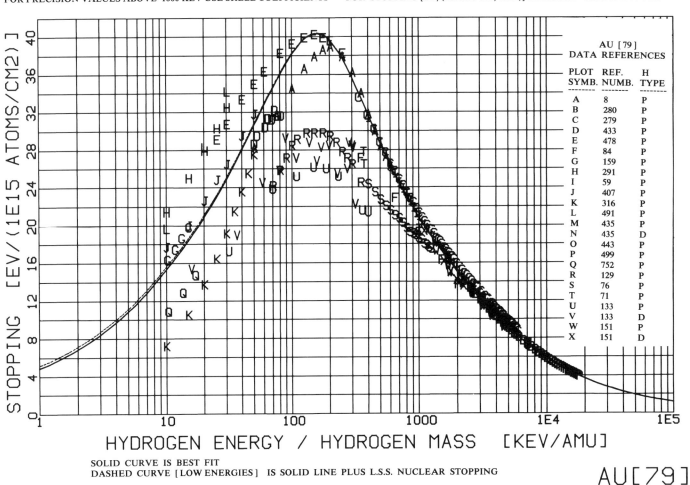

------------ENERGY IN KEV-------------
SOLID CURVE = (S[LOW]) (S[HIGH]) / (S[LOW] + S[HIGH])

$S[LOW] = 5.46 \ (ENERGY^{0.45})$
$S[HIGH] = (1.832E4 / ENERGY) \ LN \ [1+(438.5 / ENERGY) + (0.002542 \ ENERGY)]$

ENERGIES BELOW 10 KEV : STOPPING $= 4.856 \ (ENERGY^{0.5})$
FOR PRECISION VALUES ABOVE 1000 KEV USE SHELL COEFFICIENTS

PROTON MASS = 1.008 AMU
DEUTERON MASS = 2.014 AMU
TRITON MASS = 3.017 AMU

AU[79]- ATOMIC DENSITY = 5.905E22 ATOMS / CM3
MASS DENSITY = 19.31 GRAMS / CM3

FOR STOPPING [KEV/MICROMETER] MULTIPLY GRAPH BY: 5.905
FOR STOPPING [EV/(MICROGM/CM2)] MULTIPLY GRAPH BY: 3.059

AU[79]

PLOT SYMB.	REF. NUMB.	H TYPE
A	8	P
B	280	P
C	279	P
D	433	P
E	478	P
F	84	P
G	159	P
H	291	P
I	59	P
J	407	P
K	316	P
L	491	P
M	435	P
N	435	D
O	443	P
P	499	P
Q	752	P
R	129	P
S	76	P
T	71	P
U	133	P
V	133	D
W	151	P
X	151	D

AU [79]
DATA REFERENCES

STOPPING [EV/(1E15 ATOMS/CM2)]

HYDROGEN ENERGY / HYDROGEN MASS [KEV/AMU]

SOLID CURVE IS BEST FIT
DASHED CURVE [LOW ENERGIES] IS SOLID LINE PLUS L.S.S. NUCLEAR STOPPING

AU[79]

Fig. 1 A compilation of experimental hydrogen stopping power data [15] for gold targets, (10 keV ≤ E ≤ 20 MeV). The symbols refer to the sources in Table III.

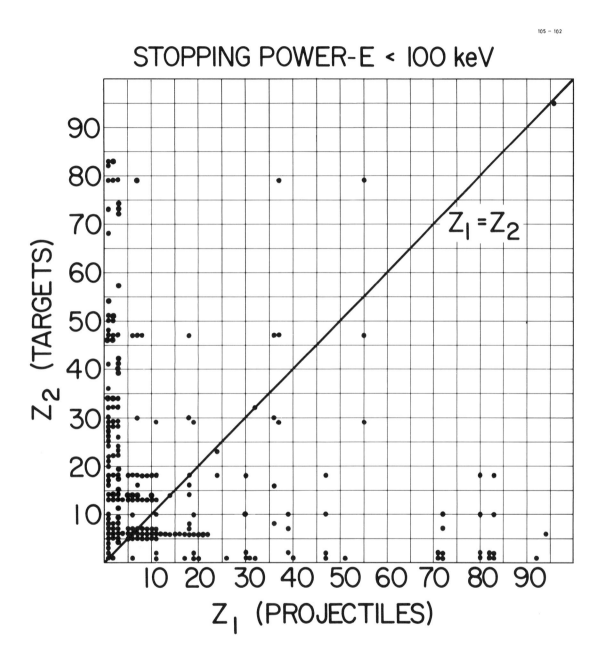

Fig. 2 Projectile-target combinations, where experimental stopping power data exist for E < 100   keV.

RANGE-E <100 keV

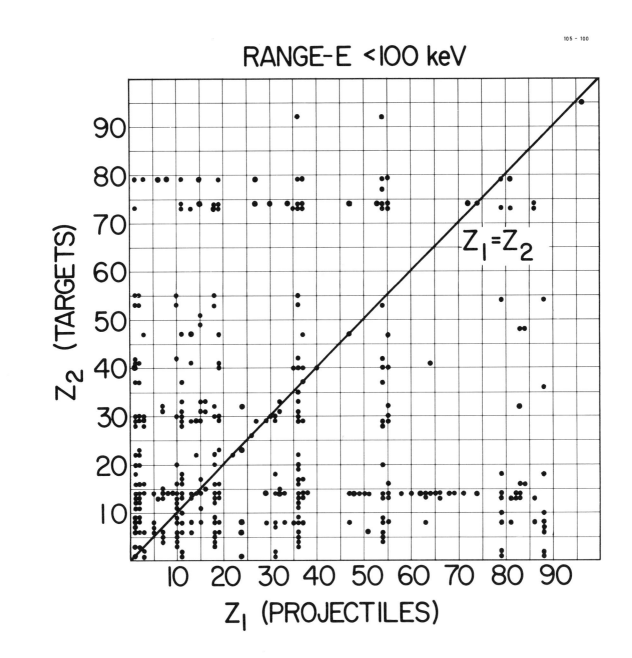

Fig. 3 Projectile-target combinations, where experimental range data exist for E < 100 keV.

105 - 104

STOPPING POWER-100 keV ≤ E < 3 MeV

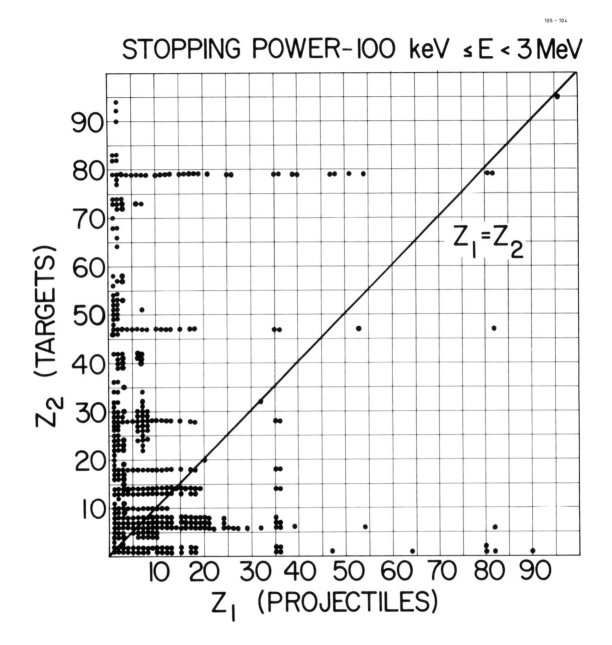

Fig. 4 Projectile-target combinations, where experimental stopping power data exist for 100 keV ≤ E ≤ 3 MeV.

12

105 - 101

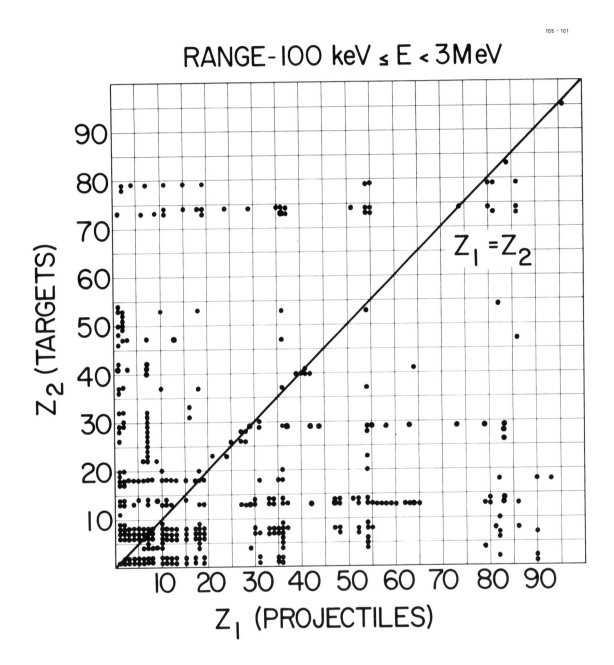

Fig. 5 Projectile-target combinations, where experimental range data exist for 100 keV $\leq$ E $<$ 3 MeV.

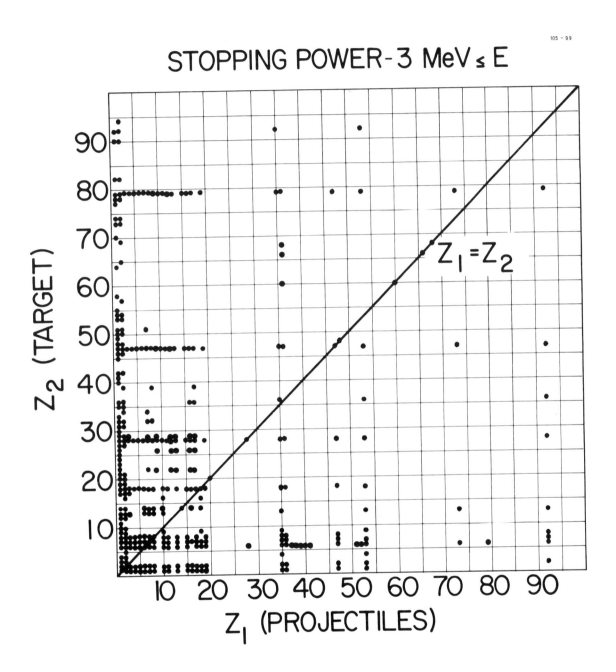

Fig. 6 Projectile-target combinations, where experimental stopping power data exist for 3 MeV ≤ E.

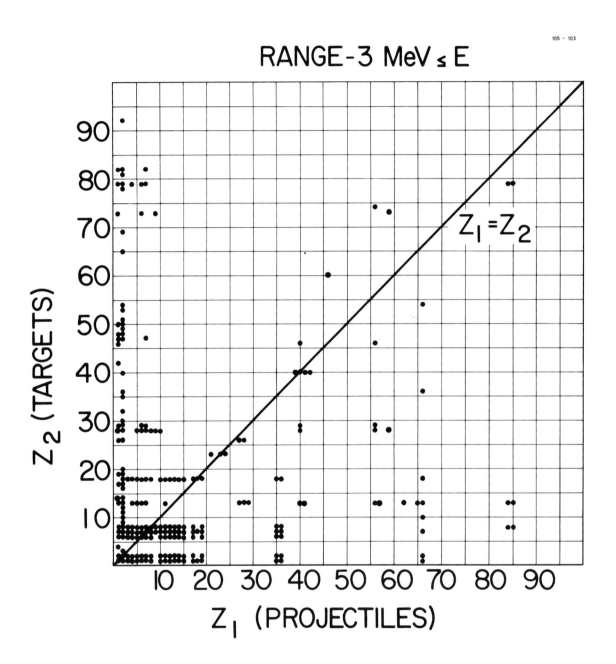

Fig. 7 Projectile-target combinations, where experimental range data exist for 3 MeV $\leq$ E.

TABLE III

TARGETS

HYDROGEN					H 1
Stopping Power			Range		
Energy	Projectiles	Ref.	Energy	Projectiles	Ref.
10-80k	p	99	20-64k	p	53
60-340k	p,d rel. to air target	40	1.5-4.5M	α	36
30-600k	p → H_2, NH_3	103	5.3-7.7M	α	64
20-700k	p		5.3M	α	90
150-400k	d, α, N, Ne	131			23
4.43M	p	26			51
5.3M	α → H_2, H_2S	65	5.3,5.5M	α	118
5.3,6.1M	α	61	5.5M	α	123
4.2-7.7M	α	85	5.5M	α → HCl, NH_3; rel. to air	97
7.7M	α rel. to air target	117	725k	Th	145
100-450k	Li	2	725k	Th	
240M	p rel. to Cu target	147	97k	Ra	144
8-40M	α		0.6-1.2M	^{66}Ga	156
24-120M	C		6-21M	Dy	208
80-400M	Ar	148	40-450k	^{8}Li	210
40-60k	He	177	96k	^{224}Ra	296
340M	p rel. to Al and Cu targets	218			446
635M	p rel. to Cu target	222			447
1.5-30k	p,d	236	50k	^{24}Na, ^{66}Ga, ^{198}Au	311
1-8M	α	244	0.03-9M	He, Li, Be, B, C, N, O, Ne, Na,	
1M	p; δS only	282		Mg, Al, P, Cl, K, Br, Kr (velocity	
5-90M	^{32}S, ^{35}Cl, ^{79}Br, ^{127}I	354		$2.6 \times 10^8 - 11.8 \times 10^8$ cm/sec)	362
0.03-9M	He, Li, Be, B, C, N, O, Ne, Na,		0.5-5M	^{7}Li	371
	Mg, Al, P, Cl, K, Br, Kr (vel.		12-40M	p	445
	$2.6 \times 10^8 - 11.8 \times 10^8$ cm/sec)	362	5.3M	α → H_2, NH_3	567
0.5-5M	^{7}Li	371	8.78M	α	568
0.6-3.75M	^{6}Li	370	4-250k	p	
100-200k	p; also δS		20-250k	α	762
1M	p		0.5-5.3M	α	790
100-300k	^{107}Ag				
100-500k	^{202}Hg; also δS	406			
100-500k	p; also δS	429			
0.5-2.0M	α	431			
0.3-2.0M	α → H_2, NH_3	439			
		440			
5.3M	α → H_2, NH_3; rel. to air	567			
8.78M	α rel. to air target	568			
100-800k	Li → H_3BO_3	605			
20-50k	^{35}Cl, ^{69}Ga				
30-50k	^{90}Zr, ^{121}Sb, ^{208}Pb				
40,50k	^{56}Fe				
50k	^{40}Ca, ^{74}Ge, ^{238}U	652			
3-17M	α → $La_2Mg_3(NO_3)_{12} \cdot 24H_2O$	661			
100-500k	Gd, Hg, Pb, Th; also δS	663			
50k	^{12}C, ^{23}Na, ^{35}Cl, ^{66}Zn, ^{89}Y,				
	^{109}Ag, ^{138}Ba, ^{160}Gd, ^{178}Hf,				
	^{198}Hg, ^{209}Bi; also δS	396			
1-5M	α	233			
100-300k	He, O; also δS	421			
40-120k	p → TiH	737			

(CONT'D)

HYDROGEN (CONT'D)

0.3-2.5M	p, $\alpha \rightarrow H_2$, H_2S	785
5M	$\alpha \rightarrow NH_3$	438
	Fiss. fragm.	259
		432
80-840k	Li $\rightarrow HBO_2$, H_3BO_3, KBH_4, $LiBH_4$, $NaBH_4$, NH_4BF_4	813
100k	Li $\rightarrow$ LiOH, HCl, H_2SO_4 (Dep. on conc.)	815
2.54-4.93M	α	828
20-156k	Pb	876

DEUTERIUM					D 1
	Stopping Power			Range	
Energy	Projectiles	Ref.	Energy	Projectiles	Ref.
40 1.5-30k	p,d	236	142,192k	p	33
50k	^{12}C, ^{23}Na, ^{35}Cl, ^{39}K, ^{66}Zn, ^{89}Y, ^{109}Ag, ^{138}Ba, ^{160}Gd, ^{175}Lu, ^{178}Hf, ^{198}Hg, ^{209}Bi; also δS		5.3M	α	45
			725k	Th	145
		396	725k	Th	
0.5-2.0M	α	431	97k	Ra	144
	Fiss. fragm.	259	0.6-1.2M	Dy	203
		432	40-450k	^8Li	210
			50k	^{24}Na, ^{66}Ga, ^{198}Au	311
			97k	^{224}Ra (target also $^3H_2$)	447
			0.05k	Au	456

HELIUM					He 2
Stopping Power			Range		
Energy	Projectiles	Ref.	Energy	Projectiles	Ref.
10-80k	p	99	50-300k	H, He, N, N_2, Ne, Ar	57
60-340k	p,d; rel. to air target	40	1.5-4.5M	α	36
30-600k	p	103	5.3,5.5M	α	118
20-700k	p		5.3M	α	23
150-450k	d		5.5M	α	123
150-400k	He, N, Ne	131	6.1M	α	10
4.43M	p	26	5.3-10.5M	α	19
5.3,6.1M	α	61	20-250k	He, N, Ne, Ar	48
4.2-7.7M	α	85	725k	Th	145
3-7M	α rel. to air targets	93	725k	Th	
100-450k	Li	2	97k	Ra	144
24-120M	C	148	0.6-1.2M	^{66}Ga	156
40-250k	H	175	6-21M	Dy	185
1M	p; δS only	282	24-120M	C	208
5-90M	^{32}S, ^{35}Cl, ^{79}Br, ^{127}I	354	40-450k	^{8}Li	210
0.03-8M	He, Li, Be, B, C, N, O, Ne, Na,		5.3M	α	567
	Mg, Al, P, Cl, K, Br, Kr (vel.		96k	^{224}Ra	296
	3.8×10^8 - 11.8×10^8 cm/sec)	362			446
0.6-3.75M	^{6}Li	370	50k	^{66}Ga	311
1-3.5M	α; also δS	388	0.03-8M	He, Li, Be, B, C, N, O, Ne, Na,	
50k	^{23}Na, ^{39}K, ^{66}Zn, ^{89}Y, ^{109}Ag,			Mg, Al, Cl, K, Br, Kr (velocity	
	^{138}Ba, ^{160}Gd, ^{175}Lu, ^{178}Hf,			3.8×10^8 - 11.8×10^8 cm/sec)	362
	^{198}Hg, ^{209}Bi; also δS	396	30.6M	^{3}H	
100-200k	H		8.43M	^{4}H	
100-600k	He		4.12M	μ^+, μ^-	418
100-500k	^{7}Li		97k	^{224}Ra $\rightarrow$ ^{3}He, ^{4}He	447
200-500k	^{11}B, N, Mg		0.05k	Au	458
200-400k	C, Ne		8.78M	α	568
150-500k	O		103k	^{206}Pb	239
100-400k	^{202}Hg; all δS also	406	0.5-5.3M	α	790
100-500k	p; δS also	429			
	Fiss. fragments	432			
5.2M	α	472			
1.3-140G	μ rel. to 1.3 GeV/c	101			
5-110M	U	522			
25-75k	Pb	524			
5.3M	α rel. to air target	567			
8.78M	α rel. to air target	568			
9G	p, $\pi \rightarrow$ 60% He, 30% Ar, 10%				
	CH_4; also δS	626			
3-5.3M	α	666			
0.3-2.0M	α	675			
100-500k	He, Li				
200-500k	Be, B, C, N, O, F, Ne, Na, Mg;				
	also δS	421			
70k	Li	776			
46M	p $\rightarrow$ (He, CO_2); δS only	401			
		392			
		425			
0.3-12G	$\mu \rightarrow$ (He,Ar) rel. to min. S	100			
0.5-50G	$\mu \rightarrow$ (He,Xe) rel. to min. S	120			
15M	Ne; also δS	792			
0.96-5.13M	α	828			

LITHIUM					Li 3
Stopping Power			Range		
Energy	Projectiles	Ref.	Energy	Projectiles	Ref.
0.075-1.4M	p		5.5M	α	123
50-600k	p → LiF	8	1-10k	He → LiF	192
440k	p	7	1-10k	H, He	
5.3-8.8M	α	110	1-30k	Ne, Ar; all → LiF	191
340M	p rel. to Al and Cu targets	218	5-60k	H, He, Ne, Ar, Kr → LiF	416
28M	d → LiF				
56M	α → LiF	571			
		134			
35-400k	p	66			
0.2-1.3M	p	727			
100k	Li → LiOH, LiCl, LiNO$_3$ (dep. on conc.)	815			
80-840k	Li → LiBH$_4$	813			
16-27.5M	d → LiF (Constr)	892			

BERYLLIUM					Be 4
Stopping Power			Range		
Energy	Projectiles	Ref.	Energy	Projectiles	Ref.
50-400k	p	129	6-18M	p	14
0.4-1.35M	p	76	50-500k	N, Ne, Ar, Kr, Xe	164
0.35-2.0M	p	84	340M	p	209
0.05-2.0M	p	8	0.5-2.0M	Ar, Kr, Xe	310
0.5-1.5M	p	83	0.4-1.9M	C	
28.7M	p rel. to Al target	146	0.5-2.0M	N, F, Ne	
19.8M	p rel. to Al target	149	0.3-2.0M	O	350
1.5-4.5M	p,d	151	100k	Au	529
340M	p rel. to Al and Cu target	218	100k	Cu	536
635M	p rel. to Cu target	222			639
10-100M	Br, I	270	30k	Kr	900
5-12M	p,d	280			
90-200M	I	289			
7-35k	p	291			
0.4-1.9M	C				
0.5-2.0M	N				
0.3-2.0M	O				
0.5-2.0M	F, Ne	350			
7-40k	p,d	399			
0.4-2.0M	He	382			
28M	α	436			
80-840k	Li	813			

BORON					B 5
Stopping Power			Range		
Energy	Projectiles	Ref.	Energy	Projectiles	Ref.
15-25k	He, Li		50-500k	N, Ne, Ar, Kr, Xe	164
15-70k	B, C, N, O, F, Ne				
15-140k	Na	266			
25-70k	p	278			
0.1-3.0M	α	281			
0.2-5.3M	Li $\to$ B_4C, B, H_3BO_3, MoB, WB	605			
100-840k	Li $\to$ CrB, CrB_2, NbB, NbB_2,				
175k	MoB, MoB_2, TaB, TaB_2, WB, W_2B_5	812			
80-840k	Li $\to$ B, AlB_2, AlB_{12}, B_4C, B_2O_3, BFO_4, B_4Si, CaB_6, CrB, CrB_2, Cr_5B_3, HBO_2, H_3BO_3, HfB_2, KBF_4, KBH_4, LaB_6, $LiBH_4$, MoB, MoB_2, $Na_2B_4O_7$, $Na_2B_4O_7 \cdot 10H_2$ (borax), $NaBH_4$, NbB, NbB_2, NH_4BF_4, TaB, TaB_2, TiB_2, VB_2, WB, W_2B_5, YB_6, ZrB_2, CeB_6	813			
80-840k	Li $\to$ CrB_2	814			
70-100k	Li				
100k	Li $\to$ TaB, TaB_2, CrB, CrB_2, NbB, NbB_2, MoB, MoB_2, TaB, TaB_2, WB, W_2B_5	815			

CARBON					C 6
Stopping Power			Range		
Energy	Projectiles	Ref.	Energy	Projectiles	Ref.
10-80k	p $\to$ CO_2	99	5.3M	$\alpha \to$ CO	90
4.43M	p $\to$ CO_2	26	5.3M	$\alpha \to$ CO, CO_2	112
5.3M	$\alpha \to$ CO_2	65	5.5M	$\alpha \to$ CO	97
240M	p rel. to Cu target	147	5.3M	$\alpha \to$ CO_2	23
10-67k	p				82
10-80k	α				90
20-70k	^6Li, ^7Li, ^{23}Na		5.5M	$\alpha \to$ CO_2 rel. to air	97
12-130k	^9Be				98
12-140k	^{11}B, ^{12}C		50-500k	N, Ne, Ar, Kr, Xe	164
15-140k	^{14}N		5.3M	α	167
20-140k	^{16}O, ^{19}F, ^{20}Ne		340M	p	209
20-130k	^{24}Mg	166	0.5-2.0M	Ar, Kr, Xe	310
25-115M	^{127}I	170	0.2-1.5M	O	
992k	p	190	0.5-2.0M	Ne	350
20-145k	Li	194	50k	Al	461
2-24M	O		8-250k	p $\to$ CO	
4-40M	Cl	195	10-250k	p $\to$ CO_2	762

(CONT'D)

CARBON (CONT'D)

Energy	Target	Ref.		Energy	Reaction	Ref.
25-80k	^{27}Al			2-11k	Li → SiC (cryst.)	767
20-130k	^{29}Si			4-20k	H$_1$, H$_2$, H$_3$, D$_1$, D$_2$, D$_3$, He → SiC	
25-130k	^{31}P					288
25-125k	^{33}S			60k	Al → SiC	449
20-125k	^{35}Cl					450
25-210k	^{40}Ar					528
25-65k	^{39}K	203		5.3M	α → CO, CO$_2$	567
30-350k	p,α	232		8.78M	α → CO, CO$_2$	568
1-8M	α	244		5-50k	H → CO$_2$	642
82-380k	^{12}C			1-20k	H, Li	
73-418k	^{14}N			2-20k	D, He, Na	
81-479k	^{16}O			3-20k	K; all → SiC; also $\eta(\varepsilon)$	682
138-473k	^{19}F			0.1-5.3M	α → CO$_2$	437
81-946k	^{20}Ne			0.5-5.3M	α → CO$_2$	790
90-898k	^{23}Na			40k	Sb → Diamond	810
135-766k	^{25}Mg			40-250k	B → Diamond	811
88-875k	^{27}Al			10-45k	Li → Diamond	809
133-780k	^{28}Si			600M	p	905
137-849k	^{31}P					
168-753k	^{32}S					
134-1133k	^{35}Cl, ^{39}K					
138-1290k	^{37}Cl, ^{40}Ar					
191-874k	^{40}Ca	247				
0.36-3.2M	O, C	249				
0.3-1.3M	α					
0.36-2.6M	N$_2$					
0.4-6.2M	Ne	250				
169k	^{208}Pb	261				
10-100M	Br, I	270				
		293				
0.4-6.0M	p	279				
90-200M	I	289				
169k	Pb	344				
0.2-1.2M	^{45}Sc					
0.2-1.0M	^{47}Ti					
0.4-1.4M	^{52}Cr					
0.4-1.1M	^{55}Mn	294				
0.3-1.5M	^{56}Fe					
0.3-1.2M	^{59}Co					
0.4-1.5M	^{65}Cu, ^{74}Ge					
0.6-1.5M	^{79}Br, ^{86}Kr					
0.5-1.4M	^{89}Y	294				
		315				
0.2-1.5M	O					
0.5-2.0M	Ne	350				
4-30k	H, He	356				
1.1M	p	369				
0.4-2.0M	He	382				
65-180k	p,α	217				
340M	p rel. to Al and Cu target	218				
635M	p rel. to Cu target	222				
10-140M	Ta	394				
30-100k	Li	395				
0.5-30k	p	410				
3-10k	^{7}Li. Dep. on scatt. angle	420				
	Fiss. fragm.	432				
		476				

(CONT'D)

30-90M	U	477			
10-46k	^{7}Li				
15-46k	^{11}B, ^{14}N				
22-46k	^{12}C, ^{18}O, ^{19}F, ^{20}Ne, ^{22}Na				
32-46k	^{31}P, ^{40}Ar	479			
50k	Ne	487			
50k	^{7}Li, ^{11}B, ^{12}C, ^{14}N, ^{16}O, ^{19}F, ^{20}Ne, ^{23}Na, ^{24}Mg, ^{31}P, ^{40}Ar	488			
0.4-1.5M	C	15			
2.9M	O$^-$				
5.8M	O$_2^-$	532			
1M	H$^-$				
2M	H$_2$	533			
5.3M	$\alpha \rightarrow$ CO, CO$_2$; rel. to air	567			
8.78M	$\alpha \rightarrow$ CO, CO$_2$; rel. to air	568			
1.01M	p; δS only	597			
100-800k	Li $\rightarrow$ B$_4$C	605			
300-700k	Xe	624			
50-250k	Ar. Dep. on exit angle	636			
1-20k	H, D, He, Li, N, Na, K	654			
50-150k	Na				
50-300k	Ar. Both dep. on exit angle	662			
3-5.3M	$\alpha \rightarrow$ CO$_2$	666			
60-300k	H				
75-150k	H$_2$				
60-100k	H$_3$; all rel. to H proj.	670			
3-5.5M	$\alpha \rightarrow$ CO$_2$; δS also	687			
150k	p	247			
1-5M	α	233			
100,200k	N, Ne, Ar, Mn, Kr, Xe	719			
90-300k	Xe	720			
460k	p; δS only	754			
0.26-4.5M	p; also δS	755			
0.6-3.75M	^{6}Li $\rightarrow$ CO$_2$	370			
46M	p $\rightarrow$ (He;CO$_2$) δS only	392			
		401			
		425			
0.4-3.4M	p $\rightarrow$ CO$_2$	403			
0.1-5.3M	$\alpha \rightarrow$ CO$_2$	437			
0.3-2.0M	$\alpha \rightarrow$ CO, CO$_2$	439			
		440			
		504			
1-5M	$\alpha \rightarrow$ UC, (U,Pu)C	469			
5.2M	$\alpha \rightarrow$ CO$_2$	472			
2M	$\alpha \rightarrow$ NbC	505			
10-15M	Ne				
5-15M	Ar; also δS	792			
0.3-5.3M	$\alpha \rightarrow$ CO$_2$	793			
80-840k	Li $\rightarrow$ B$_4$C	813			
100k	Li	815			
0.46-4.79M	p; δS only	853			
121-335k	^{40}Ca	860			
84M	Y, Zr; δS only	863			
80-100M	Kr, Rb, Sr, Y, Zr, Nb, Sb, Te; also δS	867			
25-150G	p, $\pi \rightarrow$ Ar+20% Co$_2$; also δS	868			
0.3-1M	H				
0.6-2M	H$_2$				
2.4-16M	0				
4.8-32M	0$_2$	869			
0.3-2M	α	879			
22-115M	I	909			
8.78M	α; δS only	911			
200M	Au				
110M	I				
48M	Ni; δS only	915			

NITROGEN					N 7
Stopping Power			**Range**		
Energy	Projectiles	Ref.	Energy	Projectiles	Ref.
10-80k	p	99	50-300k	H, He, N, N_2, Ne, Ar	57
30-600k	p $\rightarrow$ N_2, NO, N_2O, NH_3	103	5.3M	α	23
0.4-1.05M	p	32			82
4.43M	p	26			90
240M	p rel. to Cu target	147		$\rightarrow$ N_2, N_2O	65
0.2-2.73M	^3H	142		$\rightarrow$ N_2, N_2O, NO	112
8-40M	He		20-250k	He, N, Ne, Ar	48
24-130M	C		5.5M	α $\rightarrow$ NH_3 rel. to air	97
80-400M	Ar	148	725k	Th	145
25-50k	H, He, C, N, O, Ar	229		Th	
-	Fission fragments	259	97k	Ra	144
1M	p; δS only	282	0.2-2.73M	^3H	142
10-50k	d		0.6-1.2M	^{66}Ga	
10-90k	He		3.9M	^{18}F	156
20-180k	B, C, N, O, F, Ne		6-21M	Dy	185
5-100k	p	342	8-40M	He	
5-90M	^{32}S, ^{35}Cl, ^{79}Br, ^{127}I	354	94-120M	C	
0.6-3.75M	^6LI	370	80-400M	Ar	208
0.4-3.4M	p	403	97k	^{224}Ra	296
0.5-2.0M	α	431			446
-	Fission fragments	432	50k	^{66}Ga	311
0.1-5.3M	α	437	97k	^{224}Ra $\rightarrow$ $^{14}N_2$, $^{15}N_2$	447
5M	α $\rightarrow$ N_2, NH_3	438	0.2-2.0M	p,α	457
0.3-2.0M	α $\rightarrow$ N_2, NH_3, N_2O	439	0.3-1.0M	Ar	451
		440	5.3M	α $\rightarrow$ N_2, NO, N_2O, NH_3	567
5.2M	α	472	8.78M	α	568
1-30k	p	517	5-50k	H	642
5.3M	α $\rightarrow$ N_2, NH_3, NO, N_2O; all rel. to air target	567	0.1-5.3M	α	437
8.78M	α rel. to air target	568	7-250k	p	762
3-17M	α $\rightarrow$ $La_2Mg_3(NO_3)_{12}$ • $24H_2O$	661	260k	Ga, Cd, Te $\rightarrow$ Si_3N_4	
70k	Li	776	280k	Zn $\rightarrow$ Si_3N_4	13
0.3-2.5M	p,α	785			539
2M	α $\rightarrow$ NbN	505	20-40k	S_2 $\rightarrow$ Si_3N_4	537
10M	N		2-8M	α	901
15M	Ne				
5-15M	Ar; also δS	792			
0.3-5.3M	α	793			
100k	Li $\rightarrow$ KNO_3, $LiNO_3$, $Zn(NO_3)_2$, $La(NO_3)_3$, $Ce(NO_3)_3$ (dep. on conc.)	815			
80-840k	Li $\rightarrow$ $N H_4 B F_4$	813			
5.3M	α	65			
2M	α $\rightarrow$ $Si_3 N_4$	851			
		880			

OXYGEN — O 8

Stopping Power			Range		
Energy	Projectiles	Ref.	Energy	Projectiles	Ref.
10-80k	p	99	5.3-8.8M	α	64
30-600k	p	103	5.3M	α	51
4.43M	p	26			65
5.3,6.1M	α	61			82
240M	p rel. to Cu target	147			90
8-40M	He				112
20-100M	^{10}B		5.3,5.5M	α	118
22-110M	^{11}B		5.5M	α	123
24-120M	C		6.1M	α	10
28-140M	N		8-40M	He	
32-160M	O		20-100M	^{10}B	
38-190M	F		22-110M	^{11}B	
40-200M	Ne	220	24-120M	C	
1M	p; δS only	282	28-140M	N	
-	Fission fragments	432	32-160M	O	
0.1-5.3M	α	437	38-190M	F	220
5M	α	438	97k	^{224}Ra	296
0.3-2.0M	α	439			446
		440	97k	$^{224}Ra \rightarrow {}^{16}O_2, {}^{18}O_2$	447
		73	0.2-2.0M	p,α	457
0.5-30G	μ	74	0.3-1.0M	Ar	451
1-30k	p	517	5.3M	α	567
5.3M	α rel. to air target	567	8.78M	α	568
8.78M	α rel. to air target	568	0.1-5.3M	α	437
100-800k	Li $\rightarrow$ H_3BO_3	605	13-250k	p	762
1.0-3.4M	p	403	2-8M	α	901
70k	Li	776			
100-300k	He, O; also δS	421			
0.3-2.5M	p,α	785			
80-840k	Li $\rightarrow$ B_2O_3, BPO_4, HBO_2, H_3BO_3, $Na_2B_4O_7$, $Na_2B_4O_7 \cdot 10 H_2O$	813			
100k	Li $\rightarrow$ LiOH, KNO_3, H_2SO_4, $LiNO_3$, $Zn(NO_3)_2$, $La(NO_3)_3$, $Ce(NO_3)_3$ (dep. on conc.)	815			
5.3M	α	65			

FLUORINE — F 9

Stopping Power			Range		
Energy	Projectiles	Ref.	Energy	Projectiles	Ref.
50-600k	p $\rightarrow$ CaF_2, LiF	8	5.3M	$\alpha \rightarrow CF_4$, SF_6	60
5.3M	$\alpha \rightarrow CCl_2F_2$	65	0.1-4M	$\alpha \rightarrow SF_6$	107
0.4-6.0M	p $\rightarrow$ CaF_2	279	1-10k	He $\rightarrow$ LiF	192
20-100M	^{79}Br, $^{127}I \rightarrow UF_4$	285	1-10k	H, He	
10-200M	$^{127}I \rightarrow UF_4$	289	1-30k	Ne, Ar; all $\rightarrow$ LiF, NaF MgF_2, CaF_2	
		293			191

(CONT'D)

FLUORINE (CONT'D)

Energy	Projectiles	Ref.	Energy	Projectiles	Ref.
1M	p → 16 diff. fluorines	320	5-60k	H, He, Ne, Ar, Kr → LiF	416
1.4-1.8M	p → BaF_2, CaF_2 cryst.	422	10-80k	H, He, Ne, Ar, Kr, Xe → CaF_2	643
0.3-2.0M	α → CF_4, C_2F_6, C_3F_8, $CClF_3$, CCl_2F_2, $CHCl_2F$, $CBrF_3$	468	0.6-1.0M	p → NaF	689
28M	d				
56M	α → LiF. Distr. of energy along path	134			
		571			
0.3-2.0M	α → $CBrF_3$, $C_2Br_2F_4$, $C_2H_2F_2$	504			
300-400k	p → BaF_2, CaF cryst.; also δS	679			
3-5.5M	α → CF_4, CCl_2F_2, SF_6; also δS	687			
15M	Ne, Ar → SF_6; also δS	792			
80-840k	Li → KBF_4, NH_4BF_4	813			
100k	Li → KF (dep. on conc.)	815			
16-27.5M	d → LiF (cryst)	892			

NEON					Ne 10
	Stopping Power			Range	
Energy	Projectiles	Ref.	Energy	Projectiles	Ref.
30-600k	p	103	6.1M	α	10
0.4-1.05M	p	32	5.3M	α	112
4.43M	p	26	97k	Ar	144
5.3,6.1M	α	61	6-21M	Dy	185
4.2-7.7M	α	85	97k	^{224}Ra	296
3-7M	α rel. to air targets	93			446
50k	^{12}C, ^{23}Na, ^{39}K, ^{66}Zn, ^{89}Y, ^{109}Ag, ^{138}Ba, ^{160}Gd, ^{178}Hf, ^{198}Hg, ^{209}Bi; also δS	396	50k	^{24}Na, ^{198}Au	311
0.4-3.4M	p	403	97k	^{224}Ra → ^{20}Ne, ^{22}Ne	447
100-500k	p; also δS		0.05k	Au	458
200-300k	^{11}B		5.3M	α	567
200-400k	He, C, Ne		8.78M	α	568
200-500k	N, O, Mg; all δS only	406	5.3-7.7M	α	64
1M	p	406			
100-500k	p; also δS	429			
-	Fission fragments	432			
5.2M	α	472			
1.5-100G	μ	69			
660M	p, π; δS only	514			
5.3M	α; rel. to air targets	567			
8.78M	α; rel. to air targets	568			
0.3-2.0M	α	675			
6.62M	α; also δS	680			
100-500k	He, Li				
200-500k	Be, B, C, N, O, Ne, Na Mg; also δS	421			
0.3-30G	μ → Ne + CH_4	86			
2.09-5.29M	α	828			

SODIUM					Na 11
Stopping Power			Range		
Energy	Projectiles	Ref.	Energy	Projectiles	Ref.
4.7,6.72M	p → NaCl (cryst.); also δS	599	1-10k	H, He	
6.72M	p → NaCl (cryst.); also δS	653	1-30k	Ne, Ar → NaF	191
0.5-10.5G	μ^+, μ^- → NaI; also δS	321	40k	Kr → NaCl (cryst.)	276
6.72M	p → NaCl (cryst.); also δS	375	30-60k	^{85}Kr	
4-8M	p → NaCl (cryst.)	391	40k	^{31}P → NaCl (cryst.)	460
1.5M	p → NaCl (rand. + cryst.)	408	0.6-1.0M	p → NaF	689
31.5M	p → NaI; δS only	414	5.3M	α → NaCl (cryst.)	771
61-222M	π				
0.25-5.23G	μ → NaI	55			
80-840k	Li → $Na_2B_4O_7$, $Na_2B_4O_7 \cdot$ 10H_2O, $NaBH_4$	813			

MAGNESIUM					Mg 12
Stopping Power			Range		
Energy	Projectiles	Ref.	Energy	Projectiles	Ref.
0.4-2.0M	α	382	5.5M	α	123
3-17M	α → $La_2Mg_3(NO_3)_{12} \cdot 24H_2O$	661	1-10k	H, He → MgF_2	191
4-8M	p → MgO (cryst.)	391	40k	Kr → MgO (cryst.)	276
1-2M	α → Mg0	823			
2M	α → Mg0	851			
		880			

ALUMINUM					Al 13
Stopping Power			Range		
Energy	Projectiles	Ref.	Energy	Projectiles	Ref.
30-400k	p		30-400k	p	
30-650k	d		30-650k	d	
30-1400k	α		30-1400k	α	
750-850k	^{6}Li	133	750-850k	^{6}Li	133
50-400k	p	129	0.1-2M	p	95
200-600k	p; rel. to Au target	8	10M	p	114
0.4-1.35M	p	76	6-18M	p	14
0.35-2M	p	84	18M	p	70
4M	p; rel. to air target	136	0.7-4.45M	α	56
10M	p	114	5.5M	α	123
5.3-7.7M	α	110	3M	At	75
5.3M	α	63	5-27k	d	140
5.3M	α; rel. to air target	87			
6M	α; rel. to air target	96			

(CONT'D)

ALUMINUM (CONT'D)

Energy	Particle	No.
7.68M	α	24
		88
7.68M	α; rel. to air target	117
2.74M	^{7}Li	42
0.2-2.73M	^{3}H	142
1.5-4.5M	p,d	151
2-15k	p, d, He	159
25-115M	^{127}I	170
10-55k	^{1}H	
12-50k	^{2}H	
15-65k	He	
20-70k	^{7}Li, ^{23}Na	
15-150k	^{11}B, ^{12}C	
20-130k	N, O, Ne	
20-140k	F	203
5-12M	p,d	205
		269
65-180k	He	217
340M	p; rel. to Cu target	218
750M	p; rel. to Cu target	221
600-950k	He	
0.46-1.5M	N_2	248
0.36-3.2M	O, C	249
0.3-2.0M	He	
0.4-5.2M	N_2	
0.4-6.2M	Ne	250
10-100M	Br, I	270
2-9M	α; also δS	274
5-12M	p,d	280
10-200M	I	289
		293
7-35k	H	291
-	Fission fragments	319
20-140k	H, He	324
50-54M	α	
27M	d; also δS	329
7M	p	
14M	d	353
0.6-1.6M	p; δS only	818
1-2M	α → Al_2O_2	823
68-157k	Si, $\eta(\varepsilon)$	829
-	Fission Fragments	833
		834
0.3-1.7M	α,C,N,O; δS also	846
2M	α → Al_2O_3	851
2M	α → Al_2O_3	880
0.5-2M	α; also δS	854
5-12M	p; δS only	855
		866
6.11M	α; δS only	862
1.3M	α	
4.6M	N	878
2.4-6.8M	d	
4.6-13.6M	α	
14-20M	^{6}Li	
8-23M	^{7}Li	908
22-115M	I	909

Energy	Particle	No.
4-40M	He	
10-100M	B	
12-120M	C	
14-140M	N	
16-160M	O	
19-190M	F	
20-200M	Ne	141
1-6M	p	139
1-100k	Na	143
0.2-2.73M	^{3}H	142
2-20M	^{9}Be	150
1-3M	Ne	153
2-450k	^{222}Rn	154
0.5-240k	^{133}Xe	
0.7-2.25M	^{41}Ar	157
2-600k	^{85}Kr	158
2-40k	^{125}Xe	160
40k	^{85}Kr → Al (cryst.), Al_2O_3	163
50-500k	N, Ne, Ar, Kr, Xe	164
5-160k	^{24}Na, ^{85}Kr, ^{125}Xe → Al_2O_3	167
20-160k	^{24}Na, ^{85}Kr, ^{86}Rb, ^{125}Xe → Al (cryst.)	174
24k	^{22}Na, ^{24}Na	179
2-50k	^{137}Cs	
30k	^{24}Na, ^{86}Rb	181
0.7-60k	^{24}Na	
30k	^{42}K	180
0.5-160k	^{24}Na, ^{41}Ar, ^{85}Kr, ^{125}Xe → Al_2O_3	182
1-25k	H,D,He	193
6-21M	Dy	205
2-9k	D	206
340M	p	209
4-28M	N	211
50-110M	^{12}C, ^{14}C, ^{16}O	213
750M	p	221
100M	p	227
0.55-1.64M	^{11}C	234
3.35-17.6M	^{60}Cu, ^{61}Cu	242
0.5-4.5M	Co, Ni	254
2-12M	^{142}Sm	255
2.8-14.2M	^{126}Ba, ^{128}Ba	256
k	Ti, Sc, Cr, Fe, Mn, Ni, Co, Cu, Ge, Zr, Y, Yb, Mo, Rh, Pd, Ag, Cd, Sn, Gd, Ta, Au, Th (7 x 10^{-4} < ε < 5)	260
97M	^{95}Zr	
65M	^{140}Ba → Al (cryst.)	262
0.1-0.5M	^{24}Na, ^{42}K, ^{133}Xe → Al (cryst.)	272
140-210k	^{222}Rn	298
40k	^{133}Xe →	
40-65k	^{24}Na → Al (cryst.). Temperature dependence	304
0.5-2.0M	Kr, Xe	310
328k	^{27}Mg	332
		363
1-12M	N	366

(CONT'D)

ALUMINUM (CONT'D)

Energy	Particle/Notes	Ref.
8.78M	α; δS only	911
1.6-6.8	p	
1.6-9M	d	
2.2-18.6M	^{3}He	
3.2-17.2M	α	
14-20M	^{6}Li	
8-23M	^{7}Li	779
0.03-9M	He, Li, Be, B, C, N, O, Ne, Na, Mg, Al, P, Cl, K, Br, Kr (2.6 x 10^8 - 11.8 x 10^8 cm/sec)	362
5-13.5M	p,d	
8-20M	^{3}He, ^{4}He	374
0.4-2.0M	α	382
1.4M	p → Al (cryst.)	386
10-140M	Ta	394
30-100k	Li	395
0.6-2.4M	p	398
7-40k	H, D	399
4-60k	H	407
0.5-30k	H	410
1-9M	α	411
20,49M	p	
80M	α; δS only	412
-	Fission fragments	432
7.2M	p	435
14.4M	d	443
28M	α	436
8-19M	p; δS only	471
0.7-1.4M	p	475
30-90M	U	477
18.5,19.8M	^{20}Ne	489
2.0M	α	124
5-18M	p, d	499
1-2M	α; rel. to Au and Si	503
0.5-2.25M	α → Al_2O_3, AuAl alloy	506
0.6-2.0M	α; also δS	521
3.72,4.33M	α; rel. to air target	550
5-9M	α	580
5.3M	α; δS only	589
0.3-5M	α	598
3.9M	p → AlSb (cryst.)	601
5.3M	α; rel. to air target	618
9-25M	μ^+	635
6.8M	α	665
5.5M	α; δS only	674
300k	α	677
1-2M	α; δS only	704
0.6-2.5M	p	709
5-17M	p, d	725
1.6-6.3M	p	
1.6-9M	d	
2.2-18.6M	^{3}He	
3.2-17.2M	^{4}He	729
28.8M	α	781
8.78M	α	783
4-30k	H, He → Al_2O_3	356

Energy	Species/Notes	Ref.
1.36M	^{24}Na	332
		364
3-9k	Kr	377
	Fission fragments	419
270M	N	448
4-29M	^{149}Tb	
4-15M	At, Po	557
14.7M	p	561
1.95M	^{27}Mg	
3.81M	^{24}Na	583
5.3M	α	589
100M	p	592
40k	D	
80k	D_2	
120k	D_3	594
100k	Ba, La, Ce, Pr, Nd, Sm	606
20-80k	Kr, Xe, Cs, Dy, Au, Pb, Bi	611
20-150k	Kr	
23k	Na	
40k	K	
80k	Rb	
125k	Xe, Cs	
185k	Hg; all → Al (cryst.)	613
9-25M	μ^+	635
3.2-4.2M	^{211}At	685
179.4M	^{99}Mo	
188M	^{140}Ba	699
100k	Cs, Ba, La, Sm, Eu, Tb, Au	707
40-1000k	Na, K, Kr, Xe → Al_2O_3	357
75k	^{129}Xe	
80k	^{153}Eu, ^{197}Au →	
100k	^{205}Tl → Al, Al_2O_3	
75k	^{133}Cs → Al	715
12k	H → Al_2O_3	756
10k	Kr → Al_2O_3	759
35-120M	p	761
18-38M	α	768
39.3M	μ^+	
33.0,37.6M	π^+, π^-	775
260k	Ga, As, Cd, Te →	
200k	Se → Al_2O_3	13
		539
1-7M	p; rel. to Fe, Cu, Mo, Cd, Sn, Ta, Pd targets	111
5-80k	d	337
6-16k	H → Al_2O_3	797
68.5,140M	Ba → Al (cryst., chann., random; polycryst.)	804
10-80k	^{24}Na	
20-160k	^{41}Ar, ^{85}Kr, ^{125}Xe; all → Al_2O_3	819
150k	Mo, Gd, Bi	
200k	Ag, Cu, Se, Au	
250k	Rb	
300k	Cd, Cs	826
97M	^{95}Zr, ^{95}Nb	
140M	^{140}Ba, ^{140}La	827
60k	Au	845

(CONT'D)

28

TARGETS

ALUMINUM (CONT'D)

Energy	Projectiles	Ref.	Energy	Projectiles	Ref.
5.5-8.8M	$\alpha \to Al_2O_3$	424	20k	^{133}Cs	888
9M	d; rel. to Ni, Si, Cu, Ge, Zr, Rh, Ag, Sn, and air targets	67			891
20.6M	p; rel. to Ni, Cu, Nb, Pd, Ag, Cd, In, Ta, Pt, Au and Th targets	116	3-30M	$p \to Al, Al_2O_3$	904
12M	p; rel. to Ni, Cu, Rh, Pd, Ag, Cd, In, Ta, Pt, Au and Th targets	122	660M	p	905
28,37M	α; rel. to Cu, Ag, Ta, Bi, Th targets	77	50k	Au	914
80-840k	$Li \to Al, AlB_2, AlB_{12}$	813			
100k	Li	815			
0.3-2M	α	816			

SILICON					Si 14
	Stopping Power			Range	
Energy	Projectiles	Ref.	Energy	Projectiles	Ref.
9M	d; rel. to Al target	67	0.7-4.45M	α	56
2M	p	171	40-80k	^{125}Xe	169
0.2-3.1M	Si; $\eta(\varepsilon)$	186	5-80k	^{125}Xe	173
2.8M	$p \to Si$ (cryst.)	189	40k	$^{32}P \to Si$ (cryst.)	243
0.021-3.2M	Si; $\eta(\varepsilon)$	216	20k	$Sb \to Si$ (cryst.)	301
4.85M	$p \to Si$ (cryst.)	224	0.8-1.9M	p	312
430-560k	Si; $\eta(\varepsilon)$	245	56k	$In, Ga \to Si$ (cryst.)	348
3-11M	$p \to Si$ (cryst.)	257	200-400k	$B \to Si$ (cryst.)	376
375M	$p \to Si$ (cryst.)	275	0.15-1.8M	B	
3-11M	$p \to Si$ (cryst.)	305	1.0-1.7M	As	
4-7.6M	$p, d \to Si$ (cryst.)	308	0.5-1.7M	$p \to Si$ (cryst.)	380
45,730M	p		15k	B	459
910M	α		30-300k	$^{11}B \to Si$ (cryst. chann. and random)	456
370M	π^-; also δS	314			
1.5M	$p \to Si$ (cryst.)	320	100-300k	$p \to Si$ (cryst.)	455
100-400k	Si; $\eta(\varepsilon)$	323	40-500k	B	453
100-500k	B, C, N, O, F, Ne, Na, Mg, Al, Si, P, S, Cl, Ar, K $\to$ Si (cryst.)	339	30-600k	$^{31}P \to Si$ (cryst.)	452
5-42M	p; δS only	372	50-200k	^{11}B	
29-300M	p		200k	^{27}Al	
50-200M	π^+; δS only	385	140,280k	$^{69}Ga, ^{71}Ga$	
4-8M	$p \to Si$ (cryst.)	391	100-280k	^{31}P	
0.5-30k	p	410	280k	^{75}As	
760M	p; δS only	423	120,260k	$^{121}Sb, ^{123}Sb$	
1.6M	$p \to Si$ (cryst.)	463	240k	^{209}Bi	453
120k	p	466	10,40k	B	1
0.42-2.75M	$\alpha \to Si$ (cryst., amorph.)	474	20k	^{11}B	5
0.9-5.0M	$\alpha \to Si$ (cryst.)	480	70k	B	6
0.1-18M	$\alpha \to Si$ (cryst., amorph.)	482	120k	As	12
80-840k	$Li \to B_4Si$	813	30-75k	B	17
1.5-60k	H		100k	$^{11}B \to Si$ (amorphous)	18
2M	^{11}B	817	20-22k	$^{11}B \to Si$ (amorphous)	19
1-2M	$\alpha \to SiO_2$	823			

(CONT'D)

SILICON (CONT'D)

100-200k	p,d		40-120k	^{32}P	
250k	$\alpha \rightarrow SiO_2$; δS only	844	10-100k	^{24}Na	
2M	$\alpha \rightarrow SiO_2$; Si_3N_4	851	10-40k	^{35}S, ^{64}Cu, ^{85}Kr All $\rightarrow$ Si (cryst.)	
		880	40,120k	^{32}P $\rightarrow$ Si (amorphous)	27
90M	Kr,Rb; δS only	863	180k	Kr, also δR, $\delta R\perp$	125
2-60k	H, He, B, C, N, Ne	871			510
0.9-7.7M	α		50-300k	B	128
8-40M	^{12}C		30-900k	^{31}P $\rightarrow$ Si (cryst.)	498
5-48M	^{14}N		30-200k	B $\rightarrow$ Si (polycryst.)	
2-100M	^{15}N		70-800k	B $\rightarrow$ Si (amorphous)	520
9-20M	^{16}O		45k	As	527
8-30M	^{20}Ne		60k	B $\rightarrow$ Si (amorphous)	537
11-60M	^{22}Ne		50-150k	^{10}B $\rightarrow$ Si (cryst., amorph.)	538
12-15M	^{40}Ar; All δS only	882	50k	Ar	
100M	S; also δS	883	100,180k	Kr also δR, δR	540
1-2M	α	890	100k	^{11}B	541
8.78M	$\alpha \rightarrow SiO_2$; δS only	911	30-70k	B	542
6-30M	^{3}He		10-250k	B	543
8-40M	^{4}He	484	35-130k	As	544
24-120M	C		40k	As	545
28-140M	N		40-120k	^{32}P $\rightarrow$ Si (cryst., amorph.)	546
32-160M	O	485	70-280k	B, P	
0.3-1.7M	α	492	80-480k	As	552
2.0M	α	124	100k	H	570
0.9-5.0M	p, d, $\alpha \rightarrow$ Si (cryst., amorph.)	132	40k	^{31}P, ^{32}P	602
0.3-2M	$\alpha \rightarrow$ Si (amorph.)	497	20-80k	Kr, Xe, Cs, Dy, Au, Pb, Bi	611
1-2M	α; rel. to Al target	503	20-80k	Xe	
0.2-0.4M	p $\rightarrow$ Si (cryst.)	507	23k	Na	
70k	H		40k	K	
140k	H_2		80k	Kr, Rb	
210k	H_3	513	125k	Xe, Cs	
50-160M	p; also δS	515	185k	Hg; all $\rightarrow$ Si (cryst.)	613
8.8M	$\alpha \rightarrow$ Si (cryst.)	516	70k	B	614
1M	$\alpha \rightarrow$ Si (cryst.)	517	150-200k	As $\rightarrow$ Si (cryst.)	615
50-300k	p $\rightarrow$ Si (cryst.). Chann. to random ratio	574	200k	Na $\rightarrow$	
0.7-1.8M	p $\rightarrow$ Si (cryst.) Chann. to random ratio	577	110-400k	B $\rightarrow$ Si (cryst. chann. and random)	616
200,400k	p $\rightarrow$ Si (cryst.)	579	100-300k	B	617
4.85M	p $\rightarrow$ Si (cryst.); also δS	595	30-200k	B $\rightarrow$ Si (cryst.)	620
4.7,6.72M	p $\rightarrow$ Si (cryst.); also δS	599	50-145k	^{31}P	638
2M	p $\rightarrow$ Si (cryst.); also δS	602	50-250k	As	640
5-30k	Ar $\rightarrow$ SiO$_2$	610	280k	^{31}P	644
7.7M	α		5-50k	Rb, Cs	648
255M	Ar; δS only	628	25-125k	B	649
6.72M	p $\rightarrow$ Si (cryst.); also δS	646	1-2.5M	B	
		653	1-1.6M	N	
		654	1M	P	650
10-100k	Li	655	20-60k	B	656
1.15,1.75M	p $\rightarrow$		100k	Sb	659
5.7M	N $\rightarrow$ Si (cryst.); also δS	668	100k	B	
3M	p $\rightarrow$ Si (cryst.); also δS	693	150k	P; both $\rightarrow$ Si (cryst., chann. and random)	671
0.3.2.0M	N, Si $\rightarrow$ (cryst.); also δS, $\eta(\varepsilon)$	710	40k	Pb	684
5-8.7M	α; δS only	711	100k	B	690
730M	p				692
910M	α; δS only	730	40,100k	B	695
0.5-1.6M	p $\rightarrow$ Si (cryst.)	757	1-2.5M	B	698
0.3-3.0M	C, N, O, Ne, ^{28}Si, S, Ar; also δS, $\eta(\varepsilon)$	780	16-60k	P	702
0.2-2.0M	$\alpha \rightarrow$ SiO$_2$	786	200k	P $\rightarrow$ Si (cryst.)	705
0.3-1.7M	$\alpha \rightarrow$ SiO$_2$	492	40-120k	^{32}P $\rightarrow$ Si (cryst., amorph.)	706
1.3,6.6M	α		40k	^{72}Ga $\rightarrow$ Si (cryst.)	713
4.4M	N $\rightarrow$ Si (cryst.); also δS	799	45k	As	716

(CONT'D)

SILICON (CONT'D)

		20-140k	^{107}Ag → Si, SiO$_2$	718
		30-100k	B → Si (cryst.)	721
		50k	N, O, F	743
		25,40k	^{75}As	744
		25k	As	745
		60-200k	^{10}B, ^{11}B	746
		5-150k	^{11}B → SiO$_2$	747
		50-150k	^{10}B	748
				750
		20-150k	Xe	751
		20-80k	In	
		20k	Sb	753
		2-11k	Li → SiC (cryst.)	767
		4.5M	α	773
		4-20k	H$_1$, H$_2$, H$_3$, D$_1$, D$_2$, D$_3$, He → SiC	288
		30-100k	B	
		50-150k	P; both → SiO$_2$	349
		4.2-5.6k	Ar → SiO$_2$	417
		60k	Al → SiC	449
				450
				528
		140,280k	Zn	
		300k	As	
		150-280k	Se	
		260k	Cd	
		280k	Te; all → SiO$_2$	
		280k	Zn	
		260k	Ga, Cd, Te; all → Si$_3$N$_4$	13
				539
		40-300k	^{11}B	
		40-150k	^{75}As; both → SiO$_2$	518
		60,100k	B → SiO$_2$	
		40k	B → Si$_3$N$_4$	537
		7.5-52k	H$_2$, D$_2$, He, Ne → SiO$_2$ (cryst.)	558
		60k	Al → SiC	
		20,60k	Na → SiO$_2$	621
		1-20k	H, Li	
		2-20k	D, He, Na	
		3-20k	K; all → SiC, η(ε)	682
		60,100k	B	789
		10-100k	Cs	795
		45k	^{82}Sr, ^{128}Ba, ^{134}Ce, ^{140}Nd, ^{145}Eu, ^{149}Gd, ^{152}Tb, ^{160}Er, ^{167}Tm, ^{169}Lu → Si (cryst., chann., random)	796
		11.5k	H → SiO$_2$ (cryst., amorph.)	800
		10k	H	801
		30-800k	B → Si (amorph., polycryst.)	802
		30k	B	806
		60k	Al	807
		1.5-60k	H	817
		7.5k	H	824
		34k	B	825
		6k	B	
		15k	P; both → Si (cryst.)	831
		145,260k	^{31}P	
		80,150k	^{11}B; Both also δR$_\perp$	835

(CONT'D)

SILICON (CONT'D)

			75-250k	B; $\delta R_\perp$ only	836
			10-35k	Ge, As	837
			120k	B	839
			50-100k	$^{10}B\ ^{11}B$	841
			2-60k	^{209}Bi	
			60k	^{69}Ga	842
			10-80k	Pb	
			50-400k	Bi	
			40k	Ar, Cu, Kr, Cd, Al, Dy, W, All also $\delta R_\perp$	843
			1-40k	Bi	
			5-60k	Sb, As	
			40k	Ge	
			60k	Au	845
			20k	Pb	852
			100k	P	858
			20-80k	B	859
			100-550k	F	861
			100k	$^{152}Sm, ^{153}Eu, ^{157}Gd, ^{159}Tb, ^{164}Dy$	873
			30k	P	875
			60k	Na	886
			35k	Cu	
			40k	Cd, Xe, Dy	
			45k	Kr, W	
			10-40k	Pb	
			45-400k	Bi; All $\delta R_\perp$ only	887
			20,30k	^{133}Cs	888
			3-30M	p → Si, SiO_2	904
			5-30k	B	914

PHOSPHORUS					P 15
	Stopping Power			Range	
Energy	Projectiles	Ref.	Energy	Projectiles	Ref.
80-840k	Li → BPO_4	813	40k	^{35}S → GaP (cryst.)	22
			50k	N → GaP	743

SULPHUR					S 16
Stopping Power			Range		
Energy	Projectiles	Ref.	Energy	Projectiles	Ref.
5-80k	H		1-10k	H, He	
12-80k	N		1-30k	Ne, Ar; all → ZnS	191
10-100k	He		25k	^{209}Bi → CdS (cryst.)	496
15-90k	Ar		2.4-6k	Li, Na, K, Cs → ZnS	584
25-90k	^{86}Kr; all → ZnS:Ag	295	25k	Bi → CdS	697
0.95-5.3M	α → CdS	629	25k	^{210}Po → CdS (cryst.)	749
3-55M	α → SF$_6$; also δS	687	5.3M	α	91
0.25-3M	α → CdS	778	0.1-4M	α → SF$_6$	107
0.3-2.5M	p, α → H$_2$S	785	5.3M	α → SF$_6$	60
5.3M	α → H$_2$S	65	5.3M	α → SO$_2$	90
15M	Ne, Ar, → SF$_6$; also δS	792			
100k	Li → H$_2$SO$_4$ (dep. on conc.)	815			

CHLORINE					Cl 17
Stopping Power			Range		
Energy	Projectiles	Ref.	Energy	Projectiles	Ref.
5.3M	α → CCl$_2$F$_2$, CCl$_4$, C$_2$H$_5$Cl	65	5.5M	α	97
10-80k	p → CCl$_4$	99	5.3M	α → CHCl$_3$	91
240M	p; rel. to Cu target	147	40k	Kr → NaCl	276
4.7, 6.72M	p → NaCl, KCl (cryst.); δS also	599	3-9k	Kr → KCl	377
6.72M	p → NaCl, KCl (cryst.); δS also	653	30-60k	^{85}Kr	
3-5.5M	α → CF$_2$Cl$_2$; δS also	687	40k	^{31}P → NaCl (cryst.)	460
1-8M	α → CCl$_4$	694	1-8.9M	α → CCl$_4$ (gas., liq.)	562
6.72M	p → NaCl (cryst.)	375	0.5-15M	p → KCl	717
4-8M	p → NaCl (cryst.)	391	10k	Kr → KCl	759
1.5M	p → NcCl, KCl (cryst. chann., rand.)	408	5.3M	α → NaCl, KCl (cryst.)	771
0.3-2.0M	α → CCl$_4$, CClF$_3$, CCl$_2$F$_2$, CHCl$_2$F	468	0.43-1.58k	Na	
100k	Li → HCl, KCl, LiCl, ZnCl$_2$, LaCl$_3$, CeCl$_3$ (dep. on conc.)	815	1.58-3.04k	Rb → RbCl	791

ARGON — Ar 18

Stopping Power

Energy	Projectiles	Ref.
10-80k	p	99
30-600k	p	103
20-700k	p	131
0.4-1.05M	p	32
4.43M	p	26
150-450k	d, He, N, Ne	131
5.3, 6.1M	α	61
4.2-7.7M	α	85
3-7M	α rel. to air targets	93
100-400k	Li	2
0.2-2.73M	^{3}H	142
8-40M	He	
24-120M	C	
80-400M	Ar	148
0.03-2.9k	Ar	238
1M	p; δS only	282
10-80k	D, He	
25-160k	B, C, N, O, F, Ne	
5-80k	H	342
5-90M	^{35}S, ^{35}Cl, ^{79}Br, ^{122}I	354
0.03-8M	He, Be, Li, B, C, N, O, Ne, Na, Mg, Al, P, Cl, K, Kr, Br ($2.6 \times 10^8 - 11.8 \times 10^8$ cm/sec)	362
1-9M	α	381
1-3.5M	α; also δS	388
50k	^{23}Na, ^{55}Mn, ^{66}Zn, ^{109}Ag, ^{198}Hg, ^{209}Bi, δS also	396
0.4-3.4M	p	403
0.06-3.0M	Li, B	
0.2-10M	Ne	
0.1-7.0M	N	405
100-500k	p; δS also	
1M	p	406
10-20M	Li	
4-15M	Be	
6-20M	B	
8-25M	C, N	
10-25M	O	
10-16M	Ne	
20-27M	Ar	409
100-500k	p; δS also	429
0.5-2.0M	α	431
	Fission fragments	432
1-8M	α; δS also	464
5.2M	α	472
0.3-70G	μ, π; δS also	108
160-1800M	μ	109
0.6-1.3G	μ+; δS also	80
37M	p; δ only	535
5.3M	α; rel. to air targets	567
8.78M	α; rel. to air targets	568
1.5-16G	p, π, K → Ar + 5% CH$_4$	625
9G	p, π → Ar + 5% CH$_4$, 60% He + 30% Ar + 10% CH$_4$	626

Range

Energy	Projectiles	Ref.
0.25-3.5M	α	11
5.3-7.7M	α	64
5.3M	α	23
		35
		45
		112
6.1M	α	10
20-250k	α, N, Ne, Ar	48
50-300k	p, α, N, Ne, Ar, N$_2$	57
725k	Th	145
725k	Th	
97k	Ra	144
0.2-2.73M	^{3}H	142
0.6-1.2M	^{66}Ga	
1-1.7M	^{43}K	156
6-21M	Dy	185
8-40M	He	
24-120M	C	
80-400M	Ar	208
103k	^{206}Pb	
924k	^{239}Np	230
103k	^{206}Pb	239
96k	^{224}Ra	296
		446
50k	^{24}Na, ^{66}Ga, ^{198}Au	311
0.03-8M	He, Li, Be, B, C, N, O, Ne, Na, Mg, Al, P, Cl, K, Br, Kr ($3 \times 10^8 - 12 \times 10^8$ cm/sec)	362
0.05k	Au	458
0.2-2.0M	p, α	457
0.3-1.0M	Ar	451
5.3M	α	567
8.78M	α	568
5-50k	H	642
7-250k	p	
20-250k	α	762
2-8M	α	901

(CONT'D)

ARGON (CONT'D)

Energy	Projectiles	Ref.
1-9M	α	657
6.62M	α; also δS	680
70k	Li	776
0.3-12G	$\mu \to$ Xe + Ar, He + Ar; rel. to minimum	100
0.4-40G	μ, $\pi \to$ Ar + Ethylene; rel. to minimum	119
1.5M	Ne, Ar; also δS	792
0.3-2.0M	α	675
-	fission fragments	838
83M	Kr; δS only	863
25-150G	p, $\pi \to$ Ar + 20% CO_2; δS also	868
1.5G	μ	
80M	p	
1.5,40G	π^-; all $\to$ Ar+7%, CH_4; δS only	897
5-12G	$\pi^- \to$ Ar+7% CO_2; δS only	898
3G	$\pi^- \to$ Ar+7% CO_2; δS only	916

POTASSIUM — K 19

Stopping Power			Range		
Energy	Projectiles	Ref.	Energy	Projectiles	Ref.
6.72M	p $\to$ KBr (cryst.)	375	3-9k	Kr $\to$ KCl	377
		604	0.5-15M	p $\to$ KCl	717
1.5M	p $\to$ KCl, KBr, KI (cryst. chann., rand.)	408	10k	Kr $\to$ KCl	759
4.7, 6.72M	p $\to$ KCl, KBr (cryst. chann. rand.); also δS	599	5.3M	$\alpha \to$ KCl, KBr, KJ (all cryst.)	771
6.72M	p $\to$ KCl, KBr, (cryst. chann., rand.)	653			
80-840k	Li $\to$ KBF$_4$, KBH$_4$	813			
100k	Li $\to$ KNO$_3$, KF, KCl, KBr, KJ (dep. on conc.)	815			

CALCIUM — Ca 20

Stopping Power			Range		
Energy	Projectiles	Ref.	Energy	Projectiles	Ref.
200-600k	p; rel. to Mn target		5.3M	α	123
50-600k	p $\to$ CaF$_2$	8	1-10k	H, He	
0.4-6.0M	p $\to$ CaF$_2$	279	1-30k	Ne, Ar, all $\to$ CaF$_2$	191
1.4-1.8M	p $\to$ CaF$_2$ (cryst.)	422			
19.8M	p; rel. to Al target	149			
0.4-6.0M	p	279			
5-12M	p, d	358			
2-9M	Ca	393			
300-400k	p $\to$ CaF$_2$ (cryst.); δS also	679			
800-840k	Li $\to$ CaB$_6$	813			

SCANDIUM					Sc 21
	Stopping Power			Range	
Energy	Projectiles	Ref.	Energy	Projectiles	Ref.
5-12M 10-30k	p, d p → Sc_2O_3	358 491			

TITANIUM					Ti 22
	Stopping Power			Range	
Energy	Projectiles	Ref.	Energy	Projectiles	Ref.
28.7M	p; rel. to Al target	146		Fission fragments	419
19.8M	p; rel. to Al target	149	50k	He	607
2-15k	H, D, He	159	33k	^{45}Ti	723
5-12M	p,d	358	10k	Kr → TiO_2	759
0.4-2.0M	α	382	35k	p	764
30-100k	Li	395	4.2-5.6k	Ar → TiO_2	417
0.5-30k	p	410	800k	N	798
7.2M	p	443			808
30-100k	H, D	442			
0.5-1.75M	p	493			
2M	α	501			
0.5-1.75M	α	634			
40-120k	p → TiH	737			
28.8M	α	781			
1.0-7.4M	N	782			
800k	N	798			
		848			
80-840k	Li	813			
2.2M	p				
3.8-65M	^{19}F				
4.8-84M	^{24}Mg				
5.4-94M	^{27}Al				
6.4-112M	^{32}S				
7-122M	^{35}Cl	821			
-	Fission Fragments	834			
0.5-2M	α; δS only	865			

VANADIUM					V 23
Stopping Power			Range		
Energy	Projectiles	Ref.	Energy	Projectiles	Ref.
200-600k	p; rel. to Mn target	8	1.1-2.0M	^{54}Mn	
28.7M	p; rel. to Al target	146	2.2-3.8M	^{55}Mn	
19.8M	p; rel. to Al target	149	3.0-5.1M	^{51}Cr	
5-12M	p, d	358	3.6-4.9M	^{48}V	
0.6-2.0M	p	310	2.3-3.5M	^{47}Sc	
0.4-2.0M	α	382	3.0-3.7M	^{46}Sc	309
2.0M	α	124	0.5-2.0M	Xe	310
		510	50k	He	607
0.02-0.09k	Cr	596	55k	^{42}K $\rightarrow$ V (cryst.); R_{max}	630
10-100k	Li	655	800k	N	798
800k	N	798			808
		848	0.02-0.09k	Cr $\rightarrow$ V, V_2O_5, VOC_2O_4,	
80-840k	Li $\rightarrow$ VB$_2$	813		$V(CH(COCH_3)_2)_3$	
5-12M	p; δS only	855		$V(CH(COCH_3)_2)_2$	830
		866			

CHROMIUM					Cr 24
Stopping Power			Range		
Energy	Projectiles	Ref.	Energy	Projectiles	Ref.
200-600k	p; rel. to Mn target	8	800k	N	798
65-180k	He	217			808
5-12M	p, d	358			
0.4-2.0M	α	382			
0.6-10k	p	473			
0.5-2.0M	C, O, N	481			
10-100k	Li	655			
800k	N	798			
		848			
175k	Li $\rightarrow$ CrB, CrB$_2$	812			
80-840k	Li $\rightarrow$ CrB, CrB$_2$, Cr$_5$B$_3$	813			
80-840k	Li $\rightarrow$ CrB$_2$	814			
100k	Li $\rightarrow$ CrB, CrB$_2$	815			
0.5-2M	α; δS only	865			

MANGANESE					Mn 25
Stopping Power			Range		
Energy	Projectiles	Ref.	Energy	Projectiles	Ref.
200-600k	p; rel. to Ca, V, Cr, Fe, Co, Ni, Cu, Zn, Au targets	8	800k	N	798
0.4-1.0M	p	59			808
5-12M	p, d	358			
0.4-2.0M	α	382			
40-120k	α				
25-140k	p	389			
800k	N	798			
		848			

IRON					Fe 26
Stopping Power			**Range**		
Energy	Projectiles	Ref.	Energy	Projectiles	Ref.
200-600k	p; rel. to Mn target	8	1-7M	p	111
4M	p; rel. to air target	136	5.5M	α	123
5.3-7.7M	α	110	0.5-5.0M	Ni, Co	299
19.8M	p; rel. to Al target	149	2.2-5.2M	^{56}Co	
340M	p; rel. to Cu and Al target	218	1.4-3.1M	^{57}Co	
635M	p; rel. to Cu target	222	1.4-2.9M	^{57}Ni; all $\rightarrow$ ^{54}Fe	309
5-12M	p, d	358	219k	^{56}Mn	332
0.4-2.0M	α	382			367
25-140k	p			Fission fragments	419
40-120k	α	389	0.85-1.1G	μ^+, μ^-	462
4-8M	p $\rightarrow$ Fe (cryst.)	391	30k	^{53}Fe	723
0.5-30k	p	410	9-25M	$\mu^+ \rightarrow$ steel	635
7.2M	p	443	800k	N	798
0.5-2.0M	C, N, O	481			808
0.6-2.0M	p	490	200k	Bi $\rightarrow$ Fe (cryst., chann. and random)	
2.0M	α	124			919
0.5-2.75M	$\alpha \rightarrow$ Fe$_2$O$_3$, Fe$_3$O$_4$	506			
2.4, 3.5M	d				
5.4M	α; both $\rightarrow$ μ-metal, permalloy	619			
9-25M	$\mu^+ \rightarrow$ steel	635			
10-100k	Li	655			
28.8M	α	781			
800k	N	798			
		848			
2.2M	p				
3.8-65M	^{19}F				
4.8-84M	^{24}Mg				
5.4-94M	^{27}Al				
6.4-112M	^{32}S				
7-122M	^{35}Cl	821			
1-2M	$\alpha \rightarrow$ Fe$_3$O$_4$, α-Fe$_2$O$_3$	823			
-	Fission fragments	834			
2M	$\alpha \rightarrow$ Fe$_2$O$_3$, Fe$_3$O$_4$	851			
		880			

COBALT					Co 27
Stopping Power			**Range**		
Energy	Projectiles	Ref.	Energy	Projectiles	Ref.
200-600k	p; rel. to Mn target	8	800k	N	798
28.7M	p; rel. to Al target	146			808
19.8M	p; rel. to Al target	149			
5-12M	p, d	358			
0.4-2.0M	α	382			
25-140k	p				
40-120k	α	389			
0.5-2.0M	C, N, O	481			
0.6-2.0M	p	490			

(CONT'D)

COBALT (CONT'D)

Energy	Projectiles	Ref.			
2.0M	α	124			
2.5-3.0M	$\alpha \to$ Havar	427			
0.8-3.9M	d				
2.9-6.0M	p; both $\to$ Havar	523			
2.4, 3.5M	α				
5.4M	α; both $\to$ Havar	619			
800k	N	798			
		848			
0.5-2M	α; δS only	865			

NICKEL — Ni 28

Stopping Power			Range		
Energy	Projectiles	Ref.	Energy	Projectiles	Ref.
200-600k	p	8	5.5M	α	123
0.4-1.05M	p	32	8-29M	N	104
12M	p; rel. to Al target	122	0.2-2.73M	^{3}H	142
4M	p; rel. to Air target	136	5-30k	Xe	201
20.6M	p; rel. to Al target	116	8-40M	He	
9M	d; rel. to Al target	67	20-100M	^{10}B	
5.3-7.7M	α	110	22-110M	^{11}B	
28.7M	p; rel. to Al target	146	24-120M	C	
0.2-2.73M	^{3}H	142	28-140M	N	
19.8M	p; rel. to Al target	149	32-160M	O	
1.5-4.5M	p, d	151	38-190M	F	
25-115M	^{127}I	170	40-200M	Ne	220
2-24M	O		0.4-2.5M	N	235
4-40M	Cl	195	0.17-0.65M	^{55}Co	
8-40M	He		0.22-0.70M	^{56}Co	
20-100M	^{10}B		0.57-0.69M	^{57}Co	
22-110M	^{11}B		0.30-0.68M	^{58}Co	
24-120M	C		0.57-0.67M	^{57}Ni	309
28-140M	N		0.5-2.0M	Xe	310
32-160M	O		145k	^{58}Co	332
38-190M	F				365
40-200M	Ne	220	1-12M	N	366
5-20M	^{10}B			Fission fragments	419
7-28M	N	223	5-25k	Li, Na	676
0.4-0.95M	α		2-12M	^{13}C, ^{19}F	688
0.6-1.8M	N_2	248	179.4M	^{99}Mo	
0.36-3.2M	O, C	249	188M	^{140}Ba	699
0.3-2.0M	α		35k	p	764
0.4-5.6M	N_2		14-42k	p	772
0.4-6.2M	Ne	250	800k	N	798
5-20M	^{10}B				808
7-28M	N		100-400k	d	847
5-30M	O, S	264	37M	Pr	895
20-100M	Br, I	270	2-5.2M	p	903
2-9M	α; δS also	274	400k	d $\to$ Ni (cryst., poly-cryst.); also $\delta R_\perp$	917
10-200M	I	289			
		293			

(CONT'D)

NICKEL (CONT'D)

20-95k	p, α	317	200k	Bi → Ni (cryst., chann. and random)	919
	Fission fragments	319			
5-12M	p, d	358			
0.03-8M	He, Be, Li, B, C, N, O, Ne, Na, Mg, Al, P, Cl, K, Br, Kr (2.6×10^8-11.8×10^8 cm/sec)	362			
0.4-2.0M	α	382			
25-140k	p				
40-120k	α	389			
30-100k	Li	395			
0.5-30k	p	410			
1-9M	α	411			
	Fission fragments	432			
7.2M	p				
14.4M	d	435			
3-15M	α				
8-66M	O				
10-90M	^{35}Cl	434			
17M	p; δS only	441			
30-90M	U	477			
25-250k	p	478			
0.5-2.0M	α	481			
2.0M	α	124			
		504			
0.6-2.0M	α; also δS	521			
2.9M	O$^-$				
5.8M	O$^-_2$	532			
3.72, 4.33M	α; rel. to Air target	550			
8-30M	O, F	575			
5-9M	α	580			
2.4, 3.5M	d				
5.4M	α; both → μ-metal, Permalloy	619			
7-35M	O	627			
1-2M	α; δS only	704			
12.5M	Ni	708			
28.8M	α	781			
1.0-2.4M	N	782			
800k	N	798			
		848			
2.2M	p				
3.8-65M	^{19}F				
4.8-84M	^{24}Mg				
5.4-94M	^{27}Al				
6.4-112M	^{32}S				
7-122M	^{35}Cl	821			
	Fission fragments	834			
-	p; δS only	855			
5-12M		866			
22-115M	I	909			
8.78M	α; δS only	911			

COPPER					Cu 29
Stopping Power			Range		
Energy	Projectiles	Ref.	Energy	Projectiles	Ref.
50-400k	p	129	1-7M	p; rel. to Al target	111
50-600k	p	8	6-18M	p	14
0.4-1.0M	p	59	0.7-4.45M	α	56
0.4-1.05M	p	32	5.5M	α	123
0.4-1.35M	p	76	1.0-1.3M	^{66}Ga	156
0.35-2.0M	p	84	5-27k	D	140
4M	p; rel. to air target	136	10-150k	^{85}Kr	161
12M	p; rel. to Al target	122	70-1000k	Ga	178
20.6M	p; rel. to Al target	116	25-125k	^{85}Kr	184
9M	d; rel. to Al target	67	10-28k	^{30}Si	200
28,37M	α; rel. to Al target	77	5-30k	Xe	201
5.3-7.7M	α	110	340M	p	209
5.5M	α; rel. to air target	87	50-110M	^{12}C, ^{14}C, ^{16}O	213
7.7M	α	24	750M	p	221
		88	6-14k	K	240
2.74M	^{7}Li	42	0.5-2.0M	Xe	310
28.7M	p; rel. to Al target	146	800k	W	327
19.8M	p; rel. to Al target	149	54.4M	α	329
1.5-4.5M	p, d	151	2-12k	Li	
2-15k	p, d, He	159	4-14k	Na	359
340M	p; rel. to Al target	218	57-144M	p	379
5-12M	p, d	280	60k	^{67}Cu, ^{57}Co, ^{32}P	415
		358		Fission fragments	419
7-35k	p	291	50k	Al	461
267, 650M	p		270M	N	448
377M	d		20-100k	Xe → Cu (cryst.)	494
765M	α	313	2.4-57.5k	Cu	563
4-20k	H	318	50-150k	He	607
50-54M	α; δS only	329	55k	^{42}K → Cu (cryst.). R_{max}	630
3-20k	Li, Na, K	346	9-20k	Rb	
0.5-3.5M	α	368	11.19k	Cs	641
0.4-2.0M	α	382	20-80k	Xe → Cu (cryst.)	660
25-140k	p		179.4M	^{99}Mo	
40-120k	α	389	188M	^{140}Ba	699
30-140k	Li	395	660M	p	700
7-40k	p, d	399	25k	^{62}Cu, ^{64}Cu	723
4-60k	p	407	1-16k	H_1, H_2, H_3	765
0.5-30k	p	410	20-100k	Xe → Cu (cryst.)	766
	Fission fragments	432	35-120M	p	761
7.2M	p		800k	N	798
14.4M	d	435			808
		443	150k	Mo, Au, Bi	
28M	α	436	200k	Cu, Ta	
0.6-10k	p	473	250k	Rb	
25-250k	p	478	300k	Ru, Cs, Ce, Eu	826
0.5-2.0M	α	481	54-158k	He	857
2.0M	α	124			858
5-18M	p, d	499	20k	^{133}Cs	891
1-2M	α; rel. to Ag, Au targets	503	660M	p	905
0.5-2.25M	α → Au Cu alloys	506			
5.3M	α; rel. to air targets	618			
6-16k	He	637			
9-20k	Rb				
11-19k	Cs	641			

(CONT'D)

COPPER (CONT'D)

6.8M	α	665
5-17M	p, d	725
4-16k	H, He	752
1-16k	H_1, H_2, H_3	765
1.6-7.2M	p	
2.3-8.5M	d	
5-18M	α	
10-17M	^7Li	779
28.8M	α	781
8.78M	α	783
0.5-3.5M	$\alpha \rightarrow$ Cu Au alloys	368
5.3-7.7M	$\alpha \rightarrow$ Cu Ag alloys	110
800k	N	798
		848
100k	Li	815
0.6-1.6k	p; δS only	818
2.2M	p	
3.8-65M	^{19}F	
4.8-84M	^{24}Mg	
5.4-94M	^{27}Al	
6.4-112M	^{32}S	
7-122M	^{35}Cl	821
-	Fission fragments	834
0.5-2M	α; δS also	854
0.5-2M	α; δS only	865
1.3M	α	
4.6M	N	878
3-5.8M	d	
5-11M	α	
10-17M	^7Li	908
5-10k	H	
2-13k	H; δS only	
6-12k	He; δS only	910

ZINC					Zn 30
	Stopping Power			Range	
Energy	Projectiles	Ref.	Energy	Projectiles	Ref.
200-600 k	p; rel. to Mn target	8	5.5M	α	123
4M	p; rel. to air target	136	700-1000k	Ga	178
5.3-7.7M	α	110	0.4-2.0M	$\alpha \rightarrow$ ZnTe	632
19.8M	p; rel. to Al target	149	25k	^{63}Zn	723
5-12M	p,d	358	1-10k	H, He	
0.6-2.4M	p	398	1-30k	Ne, Ar. All $\rightarrow$ ZnS	191
0.5-2.0M	α	481	5-60k	H, He, Ne, Ar, Kr $\rightarrow$ ZnS	416
0.3-2.0M	p,$\alpha \rightarrow$ ZnTe	632	2.4-6k	Li, Na, K, Cs $\rightarrow$ ZnS	584
5-80k	H		800k	N	798
25-90k	N, ^{86}Kr				808
10-100k	He				
15-90k	Ar. All $\rightarrow$ ZnS: Ag	295			
10-100k	He, N, Ar, Kr $\rightarrow$ ZnO: Zn; rel. to H (same energy)	325			
800k	N	798			
		848			
100k	Li $\rightarrow$ $ZnCl_2$, $Zn(NO_3)_2$	815			

GALLIUM					Ga 31
Stopping Power			Range		
Energy	Projectiles	Ref.	Energy	Projectiles	Ref.
4-7.6M	p,d → GaAs (cryst.)	308	80k	^{85}Kr → GaAs (cryst.)	188
3-4M	p,d,α → GaSb (cryst.)	307	10-50k	Kr → GaAs	378
1.9-80M	p		10-40k	^{35}S	
2.5-8.0M	d		40k	^{85}Kr, ^{24}Na. All → GaAs (cryst.)	
5.4-13.9M	^{3}He. All → GaSb (cryst.)	601	40k	^{35}S → GaP (cryst.)	22
800k	N	798	100k	S$_2$ → GaAs	530
		848	20-40k	S → GaAs (cryst; chann. and random)	
					681
			50k	N → GaP	743
			800k	N	798
					808

GERMANIUM					Ge 32
Stopping Power			Range		
Energy	Projectiles	Ref.	Energy	Projectiles	Ref.
0.4-1.0M	p	59	0.7-4.45M	α	56
9M	d; rel. to Al target	67	4k	Cs	197
2-15k	H, D, He	159			198
20-100k	Ge; $\eta(\varepsilon)$	226	0.8-1.9M	p	312
		283	4-20k	^{137}Cs	559
21-997k	Ge; $\eta(\varepsilon)$	246	46-82k	Ar,N	769
4.25-7.75M	p		800k	N	798
7.63M	d; both → Ge (cryst.)	252			808
3-11M	p → Ge (cryst.)	305	1k	Bi	845
4-7.6M	p,d → Ge (cryst.)	308	5k	Cr	852
0.4-2.0M	α	382			
0.36-5.49M	p	387			
4-8M	p → Ge (cryst.)	391			
0.5-30k	p	410			
3-15M	α				
8-66M	0				
10-90M	^{35}Cl	434			
2.0M	α	124			
0.3-2.0M	α → Ge (amorphous)	497			
6.72M	p → Ge (cryst.). Also δS	653			
		654			
8.8M	α → Ge (cryst.)	516			
10-100k	Li	655			
0.254k	Ge; $\eta(\varepsilon)$	664			
1-1.8k	Ge; $\eta(\varepsilon)$	672			
1.0-7.4M	N	782			
800k	N	798			
		848			
19-36k	Ge; $\eta(\varepsilon)$	829			
10-20k	Ge; $\eta(\varepsilon)$	840			
1.35G	p, π → Ge (cryst. chann. and random)				
		874			
20-260k	p, α; also δS	896			

ARSENIC					As 33
Stopping Power			Range		
Energy	Projectiles	Ref.	Energy	Projectiles	Ref.
4-7.6M	p,d → GaAs (cryst.)	308	80k	^{85}Kr → GaAs (cryst.)	188
4.5-6.9M	p → InAs (cryst.)	601	10-50k	^{85}Kr → GaAs	378
			10-40k	^{35}S	
			40k	^{85}Kr, ^{24}Na. All → GaAs (cryst.)	22
			100k	S$_2$	530
			20-40k	S → GaAs (cryst.; chann. and random)	681

SELENIUM					Se 34
Stopping Power			Range		
Energy	Projectiles	Ref.	Energy	Projectiles	Ref.
0.4-1.0M	p	59			
1.4-10M	N				
1-9M	α	411			
0.7-1.4M	p	475			
0.3-2.0M	α	500			
10-100k	Li	655			
20-260k	p, α; also δS	896			

BROMINE					Br 35
Stopping Power			Range		
Energy	Projectiles	Ref.	Energy	Projectiles	Ref.
6.7M	p → KBr (cryst.)	375	5.3M	α → CH$_3$Br	90
1.5M	p → KBr (cryst. chann., random)	408	5.5M	α → CH$_3$Br; rel. to air	97
0.3-2.0M	α → CBrF$_3$, C$_2$H$_3$Br, C$_2$H$_5$Br	468	5.3M	α → KBr (cryst.)	771
0.3-2.0M	α → C$_2$H$_3$Br, C$_2$H$_5$Br, CBrF$_3$, C$_2$Br$_2$F$_4$	504			
4.7,6.72M	p → KBr (cryst.); Also δS	599			
6.72M	p → KBr (cryst.); Also δS	604			
		653			
100k	Li → KBr (dep. on conc. in elect.)	815			

KRYPTON					Kr 36
Stopping Power			Range		
Energy	Projectiles	Ref.	Energy	Projectiles	Ref.
10-80k	p	99	6.1M	α	10
30-600k	p	103	5.3M	α	23
0.4-1.05M	p	32			112
4.43M	p	26	0.2-2.73M	^{3}H	142
5.3,6.1M	α	61	6-21M	Dy	185
0.2-2.73M	^{3}H	142	97k	^{224}Ra	446
5-90M	^{32}S, ^{35}Cl, ^{79}Br, ^{127}I	354	5.3M	α	511
1-3.5M	α; also δS	388			567
0.4-3.4M	p	403	8.78M	α	568
0.1-1M	p; also δS	406			
0.1-0.5M	p; also δS	429			
	Fission fragments	432			
5.2M	α	472			
5-110M	U	522			
5.3M	α; rel. to air target	567			
8.78M	α; rel. to air target	568			
9G	p,π → Kr + 5% CH$_4$; also δS	626			
0.3-2M	α	675			
6.62M	α; also δS	680			
0.2-5G	μ, π → Kr + 5% CH$_4$; also δS	115			
0.3-5.3M	α	793			
3-7M	α rel. to air targets	93			

RUBIDIUM					Rb 37
Stopping Power			Range		
Energy	Projectiles	Ref.	Energy	Projectiles	Ref.
			10-80k	H, He, Ne, Ar, Kr, Xe, → RbJ	643
			0.43-1.58k	Na →	
			1.58-3.04k	Rb → RbCl	791

YTTRIUM					Y 39
Stopping Power			Range		
Energy	Projectiles	Ref.	Energy	Projectiles	Ref.
3-15M	α				
8-66M	O				
10-90M	^{35}Cl	434			
0.3-2.0M	α	500			
80-840k	Li → YB$_6$	813			

ZIRCONIUM					Zr 40
Stopping Power			**Range**		
Energy	Projectiles	Ref.	Energy	Projectiles	Ref.
9M	d; rel. to Al targets	67	40k	^{133}Xe, ^{134}Cs → ZrO$_2$	241
5-12M	p,d	404	2.2-3.lM	^{90}Mo, ^{90}Nb	
0.3-2.0M	α	500	2.7-3.2M	^{89}Zr	
80-840k	Li → ZrB$_2$	813	1.7-3.1M	^{88}Zr	
800k	N	848	2.5-3.0M	^{87}Zr	
			2.8-3.0M	^{86}Zr	
			2.6-3.0M	^{86}Y	
			1.8-3.2M	^{84}Y. All → ^{90}Zr, Natl. Zr	309
			30k	^{82}Br, ^{85}Kr, ^{86}Rb, ^{125}Xe, ^{133}Xe	
				→ ZrO$_2$	328
				Fission fragments	419
			4.5M	α	773
			800k	N	808
			1 and 4k	D	832

NIOBIUM					Nb 41
Stopping Power			**Range**		
Energy	Projectiles	Ref.	Energy	Projectiles	Ref.
20.6M	p; rel. to Al targets	116		Fission fragments	419
19.8M	p; rel. to Al targets	149	50-150k	Gd	531
0.3-2.0M	α	500	220k	Li	578
50-150k	p	508			608
2M	α → NbC, NbN	505	50k	He	607
175k	Li → NbB, NbB$_2$	812	143k	Nb	609
80-840k	Li → NbB, NbB$_2$	813	55k	^{42}K → Nb (cryst.) R$_{max}$	630
100k	Li → NbB, NbB$_2$		12-17k	Rb	
300k	C	815	16-20k	Cs	641
800k	N	848	1.5-15k	^{3}He	714
500k	d	850	143k	^{92}Nb → Nb	791
			6-16k	H → Nb$_2$O$_5$	797
			50-160k	Gd	803
			4k	He → Nb (cryst., chann., random)	805
			800k	N	808
			15k	^{3}He → Nb (cryst., chann.)	
			1.5k	^{3}He → Nb (cryst., random)	849
			15k	^{3}He → Nb (cryst.)	850

MOLYBDENUM — Mo 42

Stopping Power			Range		
Energy	Projectiles	Ref.	Energy	Projectiles	Ref.
4M	p; rel. to air targets	136	1-7M	p; rel. to Al targets	111
5.3-7.7M	α	110	5-30k	Xe	201
19.8M	p; rel. to Al targets	149		Fission fragments	419
7.2M	p	443	50k	Al	461
28M	α	436	55k	^{42}K → Mo (cryst.) R_{max}	630
1-6M	p	486	10k	Kr	678
0.5-1.75M	p	493	16k	^{91}Mo	723
2.0M	α	124	800k	N	808
		501			
0.3-2.0M	α	500			
100-800k	Li → MoB	605			
0.5-1.75M	α	634			
28.8M	α	781			
175k	Li → MoB, MoB$_2$	812			
80-840k	Li → MoB, MoB$_2$	813			
100k	Li → MoB, MoB$_2$				
300k	C	815			
-	Fission fragments	834			
800k	N	848			
5-12M	p; δS only	855			
		866			

RUTHENIUM — Ru 44

Stopping Power			Range		
Energy	Projectiles	Ref.	Energy	Projectiles	Ref.
8.78M	α; δS only	911			

RHODIUM — Rh 45

Stopping Power			Range		
Energy	Projectiles	Ref.	Energy	Projectiles	Ref.
12M	p; rel. to Al targets	122			
9M	d; rel. to Al targets	67			
19.8M	p; rel. to Al targets	149			
7.2M	p				
14.4M	d	435			

PALLADIUM					Pd 46
Stopping Power			Range		
Energy	Projectiles	Ref.	Energy	Projectiles	Ref.
5.3-7.7M	α	110	1-7M	p; rel. to Al targets	111
12M	p; rel. to Al targets	122	179.4M	^{99}Mo	
20.6M	p; rel. to Al targets	116	188M	^{140}Ba	679
19.8M	p; rel. to Al targets	149			
0.4-2.0M	α	382			
20-260k	p, α; also δS	896			

SILVER					Ag 47
Stopping Power			Range		
Energy	Projectiles	Ref.	Energy	Projectiles	Ref.
50-400k	p	129	6-18M	p	14
0.4-1.0M	p	59	0.7-4.45M	α	56
0.035-2.0M	p	84	5.5M	α	123
12M	p; rel. to Al target	122	0.4-2.5M	N	235
20.6M	p; rel. to Al target	116	6-14k	K	240
9M	d; rel. to Al target	67	133k	^{272}Rn	298
28,37M	α; rel. to Al target	77	54.4M	α	329
28.7M	p; rel. to Al target	146	2-12k	Li	
19.8M	p; rel. to Al target	149	6-14k	Na	359
5.3-7.7M	α	110	1-12M	N	366
7.7M	α	24	100-400k	^{85}Kr	415
1.5-4.5M	p,d	151		Fission fragments	419
2-15k	p, d, He	159	2.9-27.5k	Ag	563
2-24k	O		220k	Li	578
4-40M	Cl	195	55k	^{42}K $\rightarrow$ Ag (cryst.). R$_{max}$	630
340M	p; rel. to Al, Cu targets	218	14k	^{106}Ag	723
5-20M	B		50-120k	Al	886
7-28M	N	223	2-5.2M	p	903
400-950k	He	248			
0.36-3.2M	O, C	249			
0.4-1.0M	D$_2$				
0.3-2.0M	He				
0.4-3.8M	N$_2$				
0.4-6.2M	Ne	250			
169k	^{208}Pb	261			
5-20M	B				
7-28M	N				
5-30M	O, S	264			
10-100M	Br, I	270			
		293			
2-9M	α. Also δS	274			
5-12M	p, d	280			
90-200M	I	289			
		293			
7-35k	H	291			
	Fission fragments	319			

(CONT'D)

19-40M	O, S	322
50-54M	α; Also δS	329
4.5M	p	355
0.03-8M	He, Bi, Li, B, C, N, O, Na, Ne, Mg, Al, P, Cl, K, Kr, Br, $(2.6 \times 10^8 - 11.8 \times 10^8$ cm/sec)	362
0.4-2.0M	α	382
10-140M	Ta	394
4-60k	p	407
1-5.3M	α	411
	Fission fragments	432
7.2M	p	
14.4M	d	443
		435
3-15M	α	
8-66M	O	
10-90M	^{35}Cl	434
0.7-1.4M	p	475
	Fission fragments	476
30-90M	U	477
25-250k	p	478
3M	α	
21.6-60M	^{127}I → Ag (cryst.)	483
1-6M	p	486
5-18M	p, d	499
1-2M	α; rel. to Cu, Au targets	503
0.5-2.25M	α → AgAu alloys	506
3.72, 4.33M	α; rel. to air targets	550
21.6-31.3M	I → Ag (cryst.). Also δS	572
		576
22.8-40.0M	I → Ag (cryst.). Charge state dependence. Also δS	573
		631
0.3-5M	α	598
5.3M	α; rel. to air targets	618
0.5-1.75M	α	634
12-17k	Rb	
16-20k	Cs	641
0.2-40k	p, d	651
6.8M	α	665
5-16k	Li	683
2-20k	H, He, C, N, O	712
1.6-6.3M	p	
1.6-9M	d	
3.0-13.5M	^{3}He	
3.2-19.2M	^{4}He	729
4-16k	H, He	752
28.8M	α	781
1.0-7.4M	N	782
5-80k	H	
12-80k	N	
10-100k	He	
15-90k	Ar	
25-90k	^{86}Kr. All → ZnS: Ag	295
5.3-7.7M	α → AuAg, AgCu alloys	
300k	C	815

(CONT'D)

SILVER (CONT'D)

Energy	Projectiles	Ref.			
0.3-2M	α	820			
1.6-6.3M	p				
1.6-9M	d				
3-13.5M	^{3}He				
3.2-19.2M	α				
8.5-21M	^{7}Li	779			
22M	p				
3.8-65M	^{19}F				
4.8-84M	^{24}Mg				
5.4-94M	^{27}Al				
6.4-122M	^{32}S				
7-122M	^{35}Cl	821			
-	Fission fragments	833			
		834			
5-12M	p; δS only	855			
		866			
6-11M	α; δS only	862			
0.5-2M	α; δS only	865			
0.6-60M	I				
0.2-5M	Kr → Ag (cryst.)	870			
1.5M	p → Ag (cryst.)	872			
1.3M	α				
4.6M	N	878			
1.5-6.5M	p				
2.6-8.5M	d	885			
3-6.8M	d				
5-13M	α				
8.5-21M	Li	894			
20-260k	p, α; also δS	896			
2.8-6.5M	d				
5-13M	α				
8.5-21M	^{7}Li	908			

CADMIUM					Cd 48
	Stopping Power			Range	
Energy	Projectiles	Ref.	Energy	Projectiles	Ref.
12M	p; rel. to Al targets	122	5.5M	α	123
20.6M	p; rel. to Al targets	116	1-7M	p; rel. to Al targets	111
19.8M	p; rel. to Al targets	149	25k	^{210}Po → CdS (cryst.)	749
5.3-7.7M	α	110	25k	^{209}Bi → CdS (cryst.)	496
148-920k	α				697
37-230k	p; both → Cd (gasous and solid)	176			
635M	p; rel. to Cu target	222			
4.5M	p	355			
0.96-5.3M	α → CdS	628			
18M	Cd	708			
0.25-3.0M	α → CdS	778			

INDIUM					In 49
Stopping Power			Range		
Energy	Projectiles	Ref.	Energy	Projectiles	Ref.
12M	p; rel. to Al targets	122	0.7-4.45M	$\alpha \rightarrow$ InSb	56
20.6M	p; rel. to Al targets	116	40k	^{32}P $\rightarrow$ InSb (cryst. chann., random)	
19.8M	p; rel. to Al targets	149			696
4.5M	p	355			
0.4-2.0M	α	382			
0.6-2.4M	p	398			
2.0M	α	124			
4.5-6.9M	p $\rightarrow$ InAs (cryst.)				
3.4-6.7M	p $\rightarrow$ InSb (cryst.)	601			

TIN					Sn 50
Stopping Power			Range		
Energy	Projectiles	Ref.	Energy	Projectiles	Ref.
0.4-1.0M	p	59	1-7M	p; rel. to Al target	111
5.3-7.7M	α	110	5.5M	α	123
7.7M	α	24	340M	p	209
5.5M	α; rel. to air target	87	660M	p	905
7.7M	α; rel. to air target	117			
9M	d; rel. to Al target	67			
19.8M	p; rel. to Al target	149			
2-15k	H, D, He	159			
340M	p; rel. to Al, Cu targets	218			
28M	α	268			
4.5M	p	355			
0.4-2.0M	α	382			
4-60k	p	407			
7.2M	p	443			
25-250k	p	478			

ANTIMONY					Sb 51
Stopping Power			Range		
Energy	Projectiles	Ref.	Energy	Projectiles	Ref.
0.4-1.0M	p	59	0.7-4.45M	$\alpha \rightarrow$ InSb	56
0.5-30k	p	410	40k	^{32}P $\rightarrow$ InSb (cryst. chann. and random)	
0.6-2.0M	p	490			696
2.0M	α	124			
0.3-2.0M	α	500			
1-10M	N	555			
1.9-8.0M	p				
2.5-8.0M	d				
5.4-13.9M	^{3}He. All $\rightarrow$ GaSb (cryst.)				
3.4-6.7M	p $\rightarrow$ InSb (cryst.)	601			
3-4M	p, d, $\alpha \rightarrow$ GaSb (cryst.)	307			
20-260k	p, α; also δS	896			

TELLURIUM					Te 52
	Stopping Power			Range	
Energy	Projectiles	Ref.	Energy	Projectiles	Ref.
2.0M	α	124	0.4-2.0M	$\alpha \rightarrow$ ZnTe	632
0.3-2.0M	α	500			
0.3-2.0M	p, $\alpha \rightarrow$ ZnTe	632			

IODINE					I 53
	Stopping Power			Range	
Energy	Projectiles	Ref.	Energy	Projectiles	Ref.
5.3M	$\alpha \rightarrow C_2H_5I$	65	5.5M	$\alpha \rightarrow CH_3I$ rel. to air target	97
0.5-10.5G	$\mu^+, \mu^- \rightarrow$ NaJ; also δS	321	5-60k	H, He, Ne, Ar, Kr $\rightarrow$ CsJ	416
4-8M	p $\rightarrow$ CsI (cryst.)	391	10-80k	H, He, Ne, Ar, Kr Xe $\rightarrow$ RbJ	643
1.5M	p $\rightarrow$ KI (cryst.; chann., random)	408	5.3M	$\alpha \rightarrow$ KJ (cryst.)	771
31.5M	p $\rightarrow$ NaI; δS only	414			
0.3-2.0M	$\alpha \rightarrow C_2H_5I$	468			
61-222M	π				
0.25-5.73G	μ; both $\rightarrow$ NaI	55			
4M	p $\rightarrow$ CsI (cryst.)	127			
100k	Li $\rightarrow$ KJ (dep. on conc. in elect.)	815			

XENON					Xe 54
	Stopping Power			Range	
Energy	Projectiles	Ref.	Energy	Projectiles	Ref.
30-600k	p	103	0.1-3M	p	68
0.4-1.05M	p	32	6.1M	α	10
4.43M	p	26	0.2-2.73M	^3H	142
3-7M	α rel. to air targets	93	6-21M	Dy	185
5.3,6.1M	α	61	97k	^{224}Ra	446
0.2-2.73M	^3H	142	0.05k	Au	458
1-3.5M	α; also δS	388	5.3M	α	511
	Fission fragments	432			567
66-100M	p $\rightarrow$ Xe + 10% CH_4; δS only	467			
0.5-60G	μ rel. to minimum	120			
5.3M	α	511			
5.3M	α rel. to air target	567			
0.3-2.0M	α	675			
1M	p; δS only	282			
0.3-12G	$\mu \rightarrow$ Xe + Ar; rel. to minimum	100			
0.5-50G	$\mu \rightarrow$ Xe + He; rel. to minimum	120			

CESIUM					Cs 55
Stopping Power			Range		
Energy	Projectiles	Ref.	Energy	Projectiles	Ref.
4M	p → CsI (cryst.)	127			
			5-60k	H, He, Ne, Ar, Kr → CsI	416

BARIUM					Ba 56
Stopping Power			Range		
Energy	Projectiles	Ref.	Energy	Projectiles	Ref.
300-400k	p → BaF_2 (cryst.); also δS	679			
1.4-1.8M	p → BaF_2 (cryst.)	422			

LANTHANUM					La 57
Stopping Power			Range		
Energy	Projectiles	Ref.	Energy	Projectiles	Ref.
0.3-2.0M	α	500			
3-17M	α → $La_2 Mg_3 (NO_3)_{12} \cdot 24H_2O$	661			
80-840k	Li → LaB_6	813			
100k	Li → $LaCl_3$, $La(NO_3)_3$ (dep. on conc. in elect.)	815			

CERIUM					Ce 58
Stopping Power			Range		
Energy	Projectiles	Ref.	Energy	Projectiles	Ref.
1-6M	p	486			
80-840k	Li → CeB_6	813			
100k	Li → $CeCl_3$, $Ce(NO_3)_3$ (dep. on conc. in elect.)	815			

NEODYMIUM					Nd 60
	Stopping Power			Range	
Energy	Projectiles	Ref.	Energy	Projectiles	Ref.
70-250M 180-300M	Nd Kr	864	49,91M	Pd	895

GADOLINIUM					Gd 64
	Stopping Power			Range	
Energy	Projectiles	Ref.	Energy	Projectiles	Ref.
5-12M 2.0M	p, d α	404 124			

TERBIUM					Tb 65
	Stopping Power			Range	
Energy	Projectiles	Ref.	Energy	Projectiles	Ref.
50-54M	α; also δS	329	54.4M	α	329

DYSPROSIUM					Dy 66
	Stopping Power			Range	
Energy	Projectiles	Ref.	Energy	Projectiles	Ref.
0.3-2.0M 70-250M 180-300M	α Dy Kr	500 864			

ERBIUM					Er 68
Stopping Power			Range		
Energy	Projectiles	Ref.	Energy	Projectiles	Ref.
0.25-2.5M	α	502			
10-30k	p $\rightarrow$ Er_2O_3	491			
0.25-2.5M	α $\rightarrow$ Er, Er_2O_3	784			
70-250M	Er				
180-300M	Kr	864			

THULIUM					Tm 69
Stopping Power			Range		
Energy	Projectiles	Ref.	Energy	Projectiles	Ref.
50-54M	α also δS	324	54.4M	α	329

YTTERBIUM					Yb 70
Stopping Power			Range		
Energy	Projectiles	Ref.	Energy	Projectiles	Ref.
1-6M	p	486			

HAFNIUM					Hf 72
Stopping Power			Range		
Energy	Projectiles	Ref.	Energy	Projectiles	Ref.
2.0M	α	124			
80-840k	Li $\rightarrow$ HfB_2	813			

TANTALUM					Ta 73
Stopping Power			Range		
Energy	Projectiles	Ref.	Energy	Projectiles	Ref.
200-600k	p; rel. to Au target	8	1-7M	p rel. to Al target	111
4M	p; rel. to air target	136	40k	^{133}Xe → Ta$_2$O$_5$	241
12M	p; rel. to Al target	122	30k	^{82}Br, ^{85}Kr, ^{86}Rb → Ta$_2$O$_5$	328
20.6M	p; rel. to Al target	116		Fission fragments	419
28,37M	α; rel. to Al target	77	50k	Al	461
28.7M	p; rel. to Al target	146	55k	^{42}K → Ta (cryst.) R$_{max}$	630
19.8M	p; rel. to Al target	149	2-12M	^{13}C, ^{19}F	688
5-13.5M	p,d		25k	^{30}Si	200
8-20M	^{3}He, ^{4}He	374	0.5-160k	^{24}Na, ^{42}K, ^{86}Rb, ^{125}Xe, ^{134}Xe,	
5-12M	p, d	404		^{204}Tl, ^{222}Rn → Ta$_2$O$_5$	228
7.2M	p	443	20-100k	^{23}Na	758
28M	α	436	5.40k	Tl, Au → Ta$_2$O$_5$	686
1-6M	p	486	6-16k	H → Ta$_2$O$_5$	797
0.5-1.75M	p	493	10-80k	^{24}Na	
20M	α	50	20-80k	^{41}Ar	
		124	20-160k	^{85}Kr, ^{125}Xe; all → Ta$_2$O$_5$	819
0.3-2.0M	α	500	37M	Pr	895
50-150k	p → Ta, Ta$_2$O$_5$	508			
0.5-1.75M	α	634			
28.8M	α	781			
8.78M	α	783.			
0.2-20M	α → Ta$_2$O$_5$	786			
175k	Li → TaB, TaB	812			
80-840k	Li → TaB, TaB$_2$	813			
100k	Li → Ta, TaB, TaB$_2$				
300k	C	815			
5-12M	p; δS only	855			
		866			

TUNGSTEN					W 74
Stopping Power			Range		
Energy	Projectiles	Ref.	Energy	Projectiles	Ref.
28.7M	p; rel. to Al target	146	2-450k	^{222}Rn	154
19.8M	p; rel. to Al target	149	2-600k	^{85}Kr	158
340M	p; rel. to Al and Cu target	218	2-40k	^{125}Xe	160
635M	p; rel. to Cu target	222	2-200k	^{133}Xe, ^{41}Ar	162
4-8M	p → W (cryst.)	391	1.6-127k	^{187}W	165
1-6M	p		1-40k	^{125}Xe	172
1-25M	α	486	5-80k	^{125}Xe → W (cryst.)	173
0.5-1.75M	p	493	0.5-160k	^{24}Na, ^{41}Ar, ^{85}Kr, ^{125}Xe → WO$_3$	182
2.0M	α	124	0.3-160k	^{24}Na, ^{41}Ar, ^{85}Kr, ^{125}Xe, ^{133}Xe	
0.3-2.0M	α	500		→ W (cryst.)	183
0.4-1.9M	α	548	2.7k	Ar, Kr	199
100-800k	Li → WB	605	40, 125k	^{133}Xe, ^{134}Cs → W (cryst.), WO$_3$	202
6.3M	p → W (cryst.)	633	20k	^{133}Xe → W (cryst.)	263
0.5-1.75M	α	634	70k	^{85}Kr → W (cryst.)	265
1.0-7.4M	N	782	0.1-1.5M	^{24}Na, ^{32}P, ^{42}K, ^{51}Cr, ^{64}Cu, ^{82}Br,	
175k	Li → WB, W$_2$B$_5$	812		^{85}Kr, ^{86}Rb, ^{122}Sb, ^{125}Xe, ^{133}Xe,	
80-840k	Li → WB, W$_2$B$_5$	813		^{225}Rn → W (cryst.)	272

(CONT'D)

TUNGSTEN (CONT'D)

100k	Li → WB, W_2B_5	815	0.1-1.5M	^{24}Na, ^{32}P, ^{42}K, ^{85}Kr, ^{133}Xe →	
6.3M	p → W (cryst.)	889		W (cryst.)	286
6.3M	p → W (cryst., chann. to random		5k	Kr → W (cryst.)	297
	ratio)	918	70-210k	^{222}Rn	298
			0.07-1M	^{42}K → W (cryst.)	302
			40k	^{24}Na, ^{133}Xe → W (cryst.) Temp.	
				dep.	304
				Fission fragments	419
			20-150k	Xe	
			23k	Na	
			40k	K	
			125k	Cs	
			80k	Rb, Kr	
			185k	Hg; all → W (cryst.)	613
			55k	^{42}K → W (cryst.). R_{max}	630
			4-9k	Cs	701
			100M	^{140}Ba	760
			10-80k	^{24}Na	
			20-80k	^{41}Ar	
			20-160k	^{85}Kr, ^{125}Xe; all → WO_3	819
			40k	^{24}Na, ^{32}P, ^{58}Co, ^{65}Zn, ^{75}Se,	
				^{85}Kr, ^{181}Hf, ^{134}Cs, ^{110m}Ag, ^{131}I,	
				^{133}Xe, ^{125}Xe → WO_3	881
			27k	Kr → W (cryst.)	884

IRIDIUM					Ir 77
	Stopping Power			Range	
Energy	Projectiles	Ref.	Energy	Projectiles	Ref.
28.7M	p; rel. to Al targets	146			
19.8M	p; rel. to Al targets	149			
2.0M	α	124			

PLATINUM					Pt 78
Stopping Power			Range		
Energy	Projectiles	Ref.	Energy	Projectiles	Ref.
4M	p; rel. to air target	136	5.5M	α	123
5.5M	α; rel. to air target	87	3-9k	Kr	377
12M	p; rel. to Al target	122			
20.6M	p; rel. to Al target	116			
19.8M	p; rel. to Al target	149			
5.3-7.7M	α	110			
5-12M	p, d	280			
7.2M	p				
14.4M	d	435			
2.0M	α	124			
1-2M	α; also δS	509			
0.6-2.0M	α; also δS	521			
0.6-2.0M	α	786			
100-200k	p,d				
250k	α; δS only	844			

GOLD					Au 79
Stopping Power			Range		
Energy	Projectiles	Ref.	Energy	Projectiles	Ref.
30-400k	p		30-400k	p	
30-650k	d		30-650k	d	
30-1400k	α		30-1400k	α	
750-850k	^{6}Li	133	750-850k	^{6}Li	133
50-400k	p	129	6-18M	p	14
364k		71	0.7-4.45M	α	56
50-600k	p		5.5M	α	123
200-600k	p rel. to Al, Mn, Cu, Ta, Pb targets	8	3M	At	75
0.4-1.0M	p	59	5-27k	d	140
0.4-1.35M	p	76	2-20M	^{9}Be	150
5.3-7.7M	α	110	0.4-6.4M	^{15}N	152
5.5M	α rel. to air target	87	2-40k	^{125}Xe	160
7.68M	α rel. to air target	117	40k	^{133}Xe $\rightarrow$ Au (cryst.); 4.3K	187
7.68M	α	24	50-110M	^{12}C, ^{14}C, O	213
		88	70k	^{85}Kr $\rightarrow$ Au (cryst.)	237
2.74M	^{7}Li	42			272
28.7M	p rel. to Al target	146	20-80k	^{133}Xe, ^{198}Au $\rightarrow$ Au (cryst.)	287
19.8M	p rel. to Al target	149	133k	^{222}Rn	298
1.5-4.5M	p, d	151	40k	^{133}Xe $\rightarrow$ Au (cryst.).Temperature dep.	303
2-15k	H, D, He	159	40-94k	^{198}Au	
25-115M	^{127}I	170	40k	^{85}Kr, ^{24}Na $\rightarrow$ Au (cryst.). Temperature dep.	304
2-24M	O		0.37-0.62M	^{200}Tl	
4-40M	Cl	195	0.41-0.82M	^{199}Tl	
5-20M	B		0.47-1.20M	^{198}Tl	
7-28M	N	223	0.88-1.46M	^{197}Tl	
13k	D $\rightarrow$ Au (cryst.)	225	1.03-1.49M	^{196}Tl	309
0.35-0.95M	He		54.4M	α	329
0.6-1.8M	N_2	248	5-15k	Li	361
0.36-3.2M	O, C	249			

(CONT'D)

GOLD (CONT'D)

0.3-2.0M	α		1-12M	N	366
0.4-3.8M	N_2		30-60k	^{85}Kr	415
0.4-6.2M	Ne	250		Fission fragments	419
40M	$^{127}I \rightarrow$ Au (cryst.)	258	40-80k	Xe	495
169k	^{208}Pb	261	4-9M	At, Po	557
5-20M	B		6.1-15.5k	Au	563
7-28M	N		9.61-14.6M	α	565
5-30M	O, S	264	40k	D	
60M	^{127}I		80k	D_2	
3M	α. Both $\rightarrow$ Au (cryst.)	267	120k	D_3	594
28M	α	268	71k	Au	609
10-100M	Br, I	270	20-150k	Xe	
		293	20-80k	Au	
2-9M	α. Also δS	274	23k	Na	
0.4-6.0M	p	279	40k	K	
5-12M	p, d	280	80k	Kr, Rb	
15k	D, He $\rightarrow$ Au, Au (cryst.)	284	125k	Cs	
14-28k	H, D, He $\rightarrow$ Au, Au (cryst.)	290	185k	Hg. All $\rightarrow$ Au (cryst.)	613
90-200M	I	289	55k	$^{42}K \rightarrow$ Au (cryst.). R_{max}	630
		293	12-17k	Rb	
7-35k	H	291	16-20k	Cs	641
10-50k	H	316	9k	^{196}Au	723
	Fission fragments	319	4.5M	α	773
19-40M	O, S	322	71k	^{196}Au	791
50-54M	α. Also δS	329	5-80k	d	337
16-30k	D $\rightarrow$ Au (cryst.)	340	40k	$^{14}C, ^{16}O, ^{32}P, ^{58}Co, ^{204}Tl$	881
400k	p				
800k	α; both $\rightarrow$ Au (cryst.)	343			
169k	^{208}Pb				
146k	^{210}Pb				
116k	^{208}Tl; all $\rightarrow$ Au (cryst.)	351			
0.03-8M	He, Be, Li, B, C, N, O, Na, Ne, Mg, Al, P, Cl, K, Kr, Br, $(2.6\times10^8-11.8\times10^8$ cm/sec)	362			
0.5-3.5M	α	368			
3M	α				
60M	^{127}I; both $\rightarrow$ Au (cryst.)	384			
0.3-1.0M	$7 \leq Z_1 \leq 54 \rightarrow$ Au (cryst.)	390			
10-140M	Ta	394			
169k	^{208}Pb	397			
7-40k	H, D	399			
4-60k	H	407			
1-5.3M	α	411			
20.49M	p				
80M	α; both δS only	412			
92k	p $\rightarrow$ Au (cryst.). Also δS	413			
2-54k	H, D, He, Li $\rightarrow$ Au (cryst.)	426			
20-28k	D $\rightarrow$ Au (cryst.)	428			
	Fission fragments	432			
20-100k	p, α	433			
7.2M	p				
14.4M	d	435			
		443			
3-15M	α				
8-66M	O				
10-90M	^{35}Cl	434			
28M	α	436			
0.4-1.9M	α	465			
-	Fission fragments	476			
30-90M	U	477			
25-250k	p	478			

(CONT'D)

GOLD (CONT'D)

3M	α				
21.6-60M	^{127}I; both → Au (cryst.)	483			
10-30k	H	491			
15-60M	^{127}I → Au (cryst.)	21			
2.0M	α	124			
5-18M	p, d	499			
0.3-2.0M	α	500			
1-2M	α; rel. to Al, Cu, Ag targets	503			
0.5-2.25M	α → AuAg, AuCu, AuAl alloys	506			
0.6-2.0M	α. Also δS	521			
3.72,4.33M	α rel. to air target	550			
6.05-6.89M	α	566			
7-38M	O				
9-30M	S	575			
5-9M	α	580			
20-80M	^{79}Br, ^{127}I → Au (cryst.)	593			
0.3-5M	α	598			
30k	H → Au (cryst.)	600			
5.3M	α; rel. to air target	618			
7-35M	O	627			
12-17k	Rb				
16-20k	Cs	641			
6.8M	α	665			
2M	α → Au (cryst.); also δS	669			
60-300k	H				
1.6-5M	p				
1.7-8.4M	d				
4.1-15.5M	α				
13-15M	^{6}Li				
10-21M	^{7}Li	779			
0.6-1.6M	p; δS only	818			
0.3-2M	α	820			
2.2M	p				
3.8-65M	^{19}F				
4.8-84M	^{24}Mg				
6.4-112M	^{32}S				
5.4-94M	^{27}Al				
7-122M	^{35}Cl	821			
40-110k	H, He, N	822			
-	Fission fragments	833			
		834			
100-200k	p,d				
250k	α, δS only	844			
5-12M	p; δS only	855			
		866			
6.11M	α; δS only	862			
1.3M	α				
4.6M	N	878			
3-6.8M	d				
5-13M	α				
8.5-21M	Li	894			
1.6-4.6M	p				
3.4-8.4M	d				
6-12M	α				
13-15M	^{6}Li				
10-21M	^{7}Li	908			
22-115M	I	909			
8.78M	α → SiO$_2$; δS only	911			
75,150k	H$_2$				
60-100k	H$_3$; all rel. to H	670			

(CONT'D)

GOLD (CONT'D)

15,21.6M	^{127}I				
10M	O; both → Au (cryst.)	673			
1-2M	α; δS only	704			
4-16k	H, He	752			
12-70M	^{79}Br				
23-82M	^{127}I; both → Au (cryst.)	777			
28.8M	α	781			
1.0-7.4M	N	782			
0.2-2.0M	α	786			
0.5-3.5M	α → CuAu alloys	368			
5.3-7.7M	α → AuAg alloys	110			

THALLIUM					Tl 81
	Stopping Power			Range	
Energy	Projectiles	Ref.	Energy	Projectiles	Ref.
			5.5M	α	123

LEAD					Pb 82
	Stopping Power			Range	
Energy	Projectiles	Ref.	Energy	Projectiles	Ref.
50-600k	p	8	5.5M	α	123
0.4-1.0M	p	59	G	μ	196
7.68M	α; rel. to air target	117	340M	p	209
5.3-7.7M	α	110	750M	p	221
19.8M	p; rel. to Al target	149	270M	N	448
340M	p; rel. to Al, Cu targets	218	35-120M	p	761
750M	p; rel. to Cu target	221	660M	p	905
1-6M	p	486			
2.0M	α	124			
5-18M	p,d	499			
0.4-1.9M	α	548			
0.6-1.6M	p; δS only	818			

BISMUTH — Bi 83

Stopping Power			Range		
Energy	Projectiles	Ref.	Energy	Projectiles	Ref.
0.5-30k	H	410	140-280k	^{208}Pb	144
0.4-1.9M	α	548			
0.4-1.0M	p	59			
28,37M	α; rel. to Al target	77			
20-260k	p, α; also δS	896			

THORIUM — Th 90

Stopping Power			Range		
Energy	Projectiles	Ref.	Energy	Projectiles	Ref.
12M	p; rel. to Al target	122			
20.6M	p; rel. to Al target	116			
28,37M	α; rel. to Al target	77			
19.8M	p; rel. to Al target	149			
1-9M	$\alpha \rightarrow$ Th, ThO$_2$	470			

URANIUM — U 92

Stopping Power			Range		
Energy	Projectiles	Ref.	Energy	Projectiles	Ref.
340M	p; rel. to Al, Cu targets	218	750M	p	221
750M	p; rel. to Cu	221	40k	^{133}Xe $\rightarrow$ UO$_2$	241
-	Fission fragments	319	4.5M	α	773
1-5M	$\alpha \rightarrow$ U, UC, (U,Pu)C	469	40k	Kr $\rightarrow$ UO$_2$ (cryst.)	276
1-9M	$\alpha \rightarrow$ U, UO$_2$	470	40k	^{133}Xe $\rightarrow$ UO$_2$	613
5-18M	p,d	499			
20-100M	^{79}Br, ^{127}I $\rightarrow$ UF$_4$	285			
10-200M	I $\rightarrow$ UF$_4$	289			
		293			

PLUTONIUM					Pu 94
Stopping Power			Range		
Energy	Projectiles	Ref.	Energy	Projectiles	Ref.
1-5M	$\alpha \rightarrow$ (U, Pu)C	469			

AMERICIUM					Am 95
Stopping Power			Range		
Energy	Projectiles	Ref.	Energy	Projectiles	Ref.
			60-400k	^{240}Cm $\rightarrow$ ^{241}AmO$_2$	525

AIR					
Stopping Power			Range		
Energy	Projectiles	Ref.	Energy	Projectiles	Ref.
60-340k	p,d	40	50-300k	p, α, N, Ne, N$_2$, Ar	57
20-700k	p		100-500k	p	16
150-450k	d, α, N, Ne	131	20-64k	p	53
30-600k	p	103	561k	p	39
4.43M	p	26	880k	p	20
9M	d; rel. to Al target	67	1-12M	p	28
7.68M	α	88	16.0,19.1,20.5	p	137
4.2-7.7M	α	85	2.75M	^{3}H	37
100-400k	Li	2	100-500k	α	16
4-5M	α; rel. to H$_2$O target	4	480-615k	α	9
5.3M	α; rel. to H$_2$, He, N$_2$, Ar, Ar+1.5% CO$_2$, Ar+10% CO$_2$, Ar + 10% CH$_4$, Kr, Kr+1.5% CO$_2$, CO$_2$ targets	23	1.47M	α	37
			0.5-20M	α	50
			1.53-2.97M	α	54
7.68M	α; rel. to CH$_3$Br, CH$_3$I, C$_2$H$_5$Cl, CCl$_4$, C$_2$H$_5$OC$_2$H$_5$, H$_2$ targets	24	1-3M	α	62
			4.77M	α	31
5.5M	α; rel. to mica target	25	5.3M	α	35
4.5-6.1M	α; rel. to H$_2$, He, O$_2$, Ar, Ne, Kr, Xe targets	61			82
					90
					112
5.3M	α; rel. to nylon, polystyrene, mica, Al targets	63	5.3-6.7M	α	64
			4.4M	α	81
4.0-5.5M	α; rel. to Al, Cu, Ag, Sn, Pt, Au, mica targets	87	5.5M	α	98
6M	α; rel. to CH$_2$, plastics, Al targets	96	4.2,4.4M	α	138
			5.3,5.5M	α	118
7.68M	α; rel. to Au, Sn, Pb, Al, H$_2$, paper, collodion targets	117	4-6M	α; rel. to H$_2$O target	89
4M	p; rel. to Al, Cu, Fe, Mo, Ni, Pt, Ta, Zn targets	136	5.5M	α; rel. to CO, CO$_2$, CH$_3$Br, CH$_3$I, Cl$_2$, HCl, NH$_3$ targets	97
			20-250k	He, N, Ne, Ar	48

(CONT'D)

AIR (CONT'D)

Energy	Projectiles	Ref.	Energy	Projectiles	Ref.
-	Fission fragments	877	0.5-7M	^{11}B	
0.5-5.5M	α; also δS	912	0.5-5M	^{13}C	78
40-250k	H	175	0.5-5.6M	α	251
200-800k	K		0.03-8M	He, Li, Be, B, C, N, O, Ne, Na, Mg, Al, P, Cl, K, Br, Kr (3×10^8 - 11.8×10^8 cm/sec)	
300-800k	Cr	341			362
0.5-5M	^7Li	371	0.5-5M	^7Li	371
1-3.5M	α; δS only	388	0.2-2.0M	p,δ	457
0.4-3.4M	p	403	0.3-2.0M	N, Na	
400k	^{20}Ne; δS only	406	0.3-1.0M	Ar	451
5.2M	α	472	0.5-2.0M	α	50
0.2-20G	μ, π rel. to minimum	79	1.47,1.78M	α	
70-1500M	μ rel. to minimum	113	0.84,1.02M	^7Li	126
5-110M	U	522	5.3,7.7M	α	512
3.72,4.33M	α rel. to Al, Ni, Ag, Au targets	550	4.9-9.5M	N	
5.3M	α rel. to He, Ne, Ar, Kr, Xe, H_2, N_2, O_2, NH_3, CO, CO_2, NO, N_2O, CH_4, C_2H_6, C_3H_8, C_4H_{10} targets	567	3.1,3.7M	Be	554
			5.3M	α	567
8.78M	α rel. to He, Ne, Ar, Kr, H_2, N_2, O_2, CO, CO_2, CH_4, C_2H_6, C_3H_8, C_4H_{10} targets	568	8.78M	α	568
			7-250k	p	
5.3M	α rel. to Al, Cu, Ag, Au targets	618	20-250k	α	762
3-5.3M	α	666	0.1-5.0M	Li	770
100-500k	He, Li		2-8M	α	901
200-500k	Be, B, C, N, O, F, Ne, Na, Mg. Also δS	421			
0.3-5.3M	α	793			

EMULSION					
Stopping Power			Range		
Energy	Projectiles	Ref.	Energy	Projectiles	Ref.
8-40M	He		4-22M	^8Li	
20-100M	^{10}B		8-16M	^8B	207
22-110M	^{11}B		4-28M	N	212
24-120M	C		12-120M	C	
28-140M	N		14-140M	N	
32-160M	O		16-160M	O	
38-190M	F		20-200M	Ne	
40-200M	Ne	220	40-400M	Ar	214
750M	p; rel. to Cu target	221	8-40M	He	
5-24G	p	292	20-100M	^{10}B	
0.5-5M	^7Li	371	22-110M	^{11}B	
4-25M	π^+, π^-	383	24-120M	C	
92k	^{237}Np	400	28-140M	N	
0.2-1.6G	π^- rel. to minimum	34	32-160M	O	
515M	π^- rel. to minimum	38	38-190M	F	
G	π, μ ($0.79<\beta<0.94$)	49	40-200M	Ne	220
5-24G	p		750M	p	221
5G	π; both rel. to minimum	58	32-320M	^{108}Ag	
24-224M	π rel. to minimum	72	30-260M	^{80}Br	231
1.4-4G	p		3-120M	N, O	253
0.3-7G	π; both rel. to minimum	121	100-240M	F, Br, I, Ta, U	300
5G	π		0.5-5M	^7Li	371
5-24G	p; both rel. to minimum	645	20M	Σ^-	444
2.5-9M	p	740			

(CONT'D)

EMULSION (CONT'D)

			100,500M	π	
			810M	μ	43
			1.47,1.78M	α	
			0.84,1.02M	^{7}Li	126
			0.5-4.0M	α	534
			87,118,146M	α	547
			14M	p	549
			2-40M	He	
			5-100M	^{10}B	
			5.5-110M	^{11}B	
			6-120M	C	
			7-140M	N	
			8-160M	O	
			8.5-190M	F	
			10-200M	Ne	551
			4.2,4.4M	α	566
			2.2-6.3M	^{11}B	
			2.2-5.0M	^{13}C	564
			7.03-19.32M	p	
			9.02-25.22M	α	569
			6.5M	p	581
			0.2-0.9M	p	
			0.6-1.5M	d	
			1.2-2.4M	α	
			0.7-1.4M	^{6}Li	582
			21.2M	p	
			5,5.5M	d	
			2.5,5M	t	
			5,10M	^{3}He	
			5M	α	
			36.6M	μ^+	
			200-700M	π^+	585
			20.6-22.4M	^{8}Li	586
			0.75-2.0M	^{6}Li	587
			9.02-25.22M	^{7}Li	588
			2.47-14.2M	p	590
			6.47M	p	591
			12M	Σ^+, Σ^-	345
			1.2-13.1M	p	
			2.1-13.0M	α	726
			0.2-16.4M	p	
			1.1-18.9M	α	728
			1-8M	p, α	732
			5-16.3M	p	
			8-19M	α	731
			17-39M	p	733
			2-13M	p	
			5-9M		734
			10.8-12.0M	p	735
			0.2-1.5M	p	736
			0.7-3.2M	p	
			1-5.3M	α	
			0.85-1.8M	^{7}Li	
			40k	C	738
			0.5-7M	^{8}Li	739
			1-17M	p	742
			1.6-5.3M	Li	741
			2-21M	p	763
			40-112M	C	787
			3-16M	α	788

MINERALS

	Stopping Power			Range	
Energy	Projectiles	Ref.	Energy	Projectiles	Ref.
0.4-1.35M	p → mica	76	97M	^{95}Zr	
0.35-2.0M	p → mica	84	65M	^{140}Ba → mica	262
1-4M	α → mica	106	5.3M	α	589
5.3M	α → mica	63	2.7-40M	Al	
5.5M	α → mica	25	4-60M	Ar, Ca	
5.5M	α → mica rel. to air	87	5-80M	Cr	
5.3-7.7M	α → mica	110	6-85M	Ni	
7.68M	α → mica	24	8-120M	Se, Kr	
		88	11-160M	Ag. All → mica	703
5.3M	α → mica δS only	589	11.5k	H → Feldspar	800

ORGANIC MATERIALS

	Stopping Power			Range	
Energy	Projectiles	Ref.	Energy	Projectiles	Ref.
4.43M	p → CH_4	26	5.3M	α → C_2H_5OH, benzene, pyridine, n-heptane, iso-octane, 2.2.4-trimethyl pentane	3
5.3M	α → polystyrene rel. to acetylene	46	5.3M	α → Ar + 10% CH_4, CH_4 rel. to air target	23
30-50k	p → celluloid	44	5.3M	α → CF_4 rel. to air target	60
5.3M	α → nylon, polystyrene; rel. to air	63	5.3M	α → CH_4	45
5.3M	α → C_3H_8, C_3H_6, C_4H_{10}, C_4H_8, CCl_4, C_2H_5Cl, CCl_2F_2, C_2H_5I	65	5.3M	α → C_2H_5OH, ether, benzene, aniline, glycerin, chloroform	91
6M	α → CH_2, polyvinyl alcohol, cellulose acetate, mylar rel. to air target	96	5.3M	α → CH_4, CH_3Br	90
10-80k	p → CCl_4	99	5.5M	α → CH_3Br, CH_3I rel. to air target	97
30-55k	p → Celluloid	102	5.5M	α → C_2H_5OH, C_6H_6, C_5H_5N	98
30-600k	p → CH_4, C_2H_5, C_2H_4, C_6H_6	103	5.3M	α → CH_4, C_2H_4	112
40-250k	p → CH_4, C_2H_2, C_2H_4, C_3H_8, $(CH_2)_3$, C_3H_6		24-120M	C → CH_4	208
40-200k	He → C_2H_4, C_3H_8	175	96k	^{224}Ra → CH_4	296
40-200k	He → CH_4, C_2H_2, C_2H_4, C_3H_8, C_3H_6, $(CH_2)_3$	215	0.03-8M	He, Li, Be, B, C, N, O, Ne, Na, Mg, Al, P, Cl, K, Br, Kr ($3 \cdot 10^8$-$12 \cdot 10^8$ cm/sec) → CH_4, benzene	362
40-340k	p, α → plastics	219	258M	Ne	
		232	50-270M	N → polyethylene	
1-8M	α → hydrocarbons	244	44-270M	N	
-	Fission fragments → mylar	259	284M	Ne, Ar → polymethylmetacrylat	448
28M	α → mylar	268	5.3M	α → 21 hydrocarbons	560
100-900k	p,d → plastics	273	1-8.9M	α → C_2H_5OH, CCl_4 (both liq. and gas. phase)	562
0.2-5.5M	α → hydrocarbons	277	5.3M	α → CH_4, C_2H_6, C_3H_8, C_4H_{10}	567
-	Fission fragments → mylar	306	8.78M	α → CH_4, C_2H_6, C_3H_8, C_4H_{10}	568
1M	p → fluorines	320		Fission fragments → makrofol, cellulose nitrate	622
15-95M	S, Cl		9-25M	μ → polythene	635
30-90M	Br		0.1-5.3M	α → CH_4	437
60-105M	I → mylar	347			
0.3-120G	$μ^+$ → scintillator plastics	352			
		360			
10-120k	$1 \le Z_1 \le 22$ → CH_4	373			
0.4-3.4M	p	403			
0.1-5.3M	α → CH_4	437			

(CONT'D)

ORGANIC MATERIALS (CONT'D)

5M	$\alpha \rightarrow$ Hydrocarbons	438
0.3-2.0M	$\alpha \rightarrow CH_4, C_2H_2, C_2H_4, C_2H_6, C_3H_6, (CH_3)_2$	439 440
0.3-2.0M	$\alpha \rightarrow CF_4, C_2F_6, C_3F_8, CCl_4, CClF_3, CCl_2F_2, CHCl_2F, CBrF_3, C_2H_3Br, C_2H_5Br, C_2H_5I$	468
5.2M	$\alpha \rightarrow CH_4$	472
0.39-2.2G	$\mu^+ \rightarrow$ xylene rel. to min.	52
0.4-100G	$\mu^+ \rightarrow$ Plastics scintillators	68
0.3-30G	$\mu \rightarrow Ne+CH_4$	86
0.3-0.8G	p	
0.3-2.2G	$\mu \rightarrow$ Plastic scintillators; Also δS	105
0.2-5G	$\mu, \pi \rightarrow Kr + CH_4$	115
0.4-40G	$\mu, \pi \rightarrow$ Ar + Ethylene rel. to minimum	119
0.3-2.0M	$\alpha \rightarrow C_2H_3Br, C_2H_5Br, CBrF_3, C_2Br_2F_4, (CH_3)_2O, C_2H_2F_2,$ many hydrocarbons	504
1.5-8M	$\alpha \rightarrow$ Many organic compounds (sol., liq. and gas phase)	526
37M	$p \rightarrow$ plast. scint.; δS only	535
5.3M	$\alpha \rightarrow CH_4, C_2H_6, C_3H_8, C_4H_{10}$ rel. to air	567
8.78M	$\alpha \rightarrow CH_4, C_2H_6, C_3H_8, C_4H_{10},$ rel. to air	568
0.3-5M	$\alpha \rightarrow$ Zapon, Paraffine	598
0.6-70k	H, He	
2-120k	$^6Li, \, ^7Li$	
3-120k	Be, B, C, N, O, F, Ne; all $\rightarrow CH_4$	612
5.4M	$\alpha \rightarrow$ Teflon	619
13-20G	$p \rightarrow$ plast. scint.	623
9G	$p, \pi \rightarrow$ Ar, He, Kr; all + CH_4; also δS	626
6.8M	$\alpha \rightarrow$ melinex	665
600M	$p \rightarrow$ plastics; δS only	667
3-5.5M	$\alpha \rightarrow CH_4, CF_4, CF_2Cl_2$; also δS	687
1-5.5M	$\alpha \rightarrow$ TEP	691
1-8M	$\alpha \rightarrow$ Liq. and sol. hydrocarbons, CCl_4	694
1-5M	$\alpha \rightarrow C_2H_6, C_4H_{10}, C_2H_4, C_3H_6, C_4H_8, C_2H_2, C_3H_4, C_4H_6, CH_4, C_6H_6,$ ethylene, polyethylene, propylene, polypropylene	233
12-120M	C	
16-160M	O	
20-200M	Ne. All $\rightarrow$ mylar, polythene	722
169k	Pb $\rightarrow$ formvar	724
0.3-5.3M	$\alpha \rightarrow CH_4$	793
0.6-1.6M	$p \rightarrow$ melinex; δD only	818
0.5-5.11M	$\alpha \rightarrow C_2H_6$	
0.59-4.44M	$\alpha \rightarrow C_3H_8$	
0.82-4.22M	$\alpha \rightarrow C_4H_{10}$	828
6.06,8.78M	$\alpha \rightarrow$ anthracene	893
1.5G	μ	
80M	p	
1.5,40G	π^-; all $\rightarrow$ Ar + 7%CH_4; δS only	797
0.3-2M	$\alpha \rightarrow C_2H_2F_2, C_2H_4F_2, C_3H_8, C_4H_6, C_4H_{10}, C_3H_4$	899

5-50k	H $\rightarrow CH_4, C_2H_5, C_2H_4, C_3H_8, C_4H_{10}$	
5-30k	N $\rightarrow CH_4$	642
130k	$^{11}C \rightarrow$ polystyrene	
85k	$^{18}F \rightarrow$ teflon	
45k	$^{34}Cl \rightarrow$ saran	723
10-250k	$p \rightarrow CH_4$	762
0.5-5.3M	$\alpha \rightarrow CH_4$	790
0.02-0.09K	Cr $\rightarrow VOC_2O_4,$ $V(CH(COCH_3)_2)_3,$ $V(CH(COCH_3)_2)_2$	830
3-30M	$p \rightarrow C_3H_5O_2$	904

(CONT'D)

ORGANIC MATERIALS (CONT'D)

50-150k	H → CH_4, CHCH, CH_2CH_2, CH_3CHCH_2, $CH_2CH_2CH_2$, CH_3CH_2 CH_3; also δS	902
	same, δS only	913
0.2-8.5M	α → polycarbonate	906
0.5-7.5M	α → polycarbonate	907
0.5-5.5M	α → (73.7% H_2+25% CO_2 + 1.3% N_2), (63.4% CH_4 + 33.4% CO_2 + 3% N_2),Melinex, mylar; also δS	912

OXIDES

Stopping Power			Range		
Energy	Projectiles	Ref.	Energy	Projectiles	Ref.
10-80k	p → CO_2	99	5.3M	α → CO_2	82
4.43M	p → CO_2	26			98
5.3M	α → CO_2, N_2O	65			23
30-600k	p → N_2O, NO	103		α → CO, CO_2, SO_2	90
10-100k	He, N, Ar, Kr rel. to H → ZnO:Zn	325		→ CO, CO_2, N_2O, NO	112
4-30k	H, He → Al_2O_3	356		α → CO_2 rel. to air	97
0.6-3.75M	^6Li → CO_2	370	40k	^{85}Kr → WO_3	162
4-8M	p → MgO	391	40k	Kr → MgO, SiO_2, UO_2 (all cryst.)	276
0.4-3.4M	p → CO_2	403	30-100k	B	
5.5-8.8M	α → Al_2O_3	424	50-150k	P → SiO_2	349
46.4M	p → He + CO_2; δS only	425	4.2-5.6k	Ar → SiO_2, TiO_2	417
0.1-5.3M	α → CO_2	437	140,280k	Zn	
5M	α → CO, CO_2	438	300k	As	
0.3-2.0M	α → N_2O, CO, CO_2	439	150-280k	Se	
		440	260k	Cd	
1-9M	α → ThO_2, UO_2	470	280k	Te → SiO_2	
5.2M	α → CO_2	472	260k	Zn, Ga, As, Cd, Te	
10-30k	Er_2O_3, Sc_2O_3	491	200k	Se → Al_2O_3	13
0.3-1.7M	α → SiO_2	492			539
0.3-2.0M	α → CO, CO_2	504	40-300k	^{11}B	
0.5-2.25M	α → Fe_2O_3, Fe_3O_4, Al_2O_2	506	40-150k	^{75}As → SiO_2	518
50-150k	p → Ta_2O_5	508	60-400k	^{240}Cm → $^{241}AmO_2$	525
5-30k	Ar → SiO_2	610	60,100k	B → SiO_2	537
3-5.3M	α → CO_2	653	7.5-52k	H_2, D_2, He, Ne → SiO_2 (cryst.)	558
5.3M	α → CO, CO_2, NO, N_2O rel. to air	567	5.3M	α → CO, CO_2, NO, N_2O	567
8.78M	α → CO, CO_2 rel. to air	568	8.78M	α → CO, CO_2	568
3-5.5M	α → CO_2; also δS	687	0.5-100k	^{125}Xe → Ta_2O_5	
0.25-2.5M	p, α → Er_2O_3	784	40k	^{133}Xe → UO_2	613
0.2-2.0M	α → SiO_2, Ta_2O_5	786	20,60k	Na → SiO_2	621
3-17M	α → $La_2Mg_3(NO_3)_{12} \cdot 24H_2O$	661	5-50k	H → CO_2	642
0.3-5.3M	α → CO_2	793	5,40k	Tl, Au → Ta_2O_5	686
1-2M	α→MgO,,Al_2O_3,SiO_2, Fe_3O_4,α-Fe_2O_3	823	0.5-160k	^{24}Na, ^{42}K, ^{86}Rb, ^{125}Xe, ^{134}Cs, ^{204}Tl, ^{222}Rn → Ta_2O_5	228
100-200k	p,d		40-1000k	Na, K, Kr, Xe → Al_2O_3	357
250k	α→SiO_2; δS only	844	0.1-5.3M	α → CO_2	437
0.3-1.7M	α,C,N,O→Al_2O_3; also δS	846	20-140k	^{107}Ag → SiO_2	718
2M	α → MgO, Al_2O_3, SiO_2, Fe_2O_3, Fe_3O_4	851	75k	^{129}Xe	
		880	80k	^{153}Eu, ^{197}Au	
			100k	^{205}Tl → Al_2O_3	715
			5-150k	^{11}B → SiO_2	747

(CONT'D)

OXIDES (CONT'D)

Energy	Projectiles	Ref.	Energy	Projectiles	Ref.
25-150G	$p, \pi \rightarrow Ar + 20\% CO_2$; also δS	868	12k	$H \rightarrow Al_2O_3$	756
5-12G	$\pi^- \rightarrow Ar + 7\% CO_2$	898	10k	$Kr \rightarrow TiO_2, Al_2O_3$	759
8.78M	$\alpha \rightarrow SiO_2$; δS only	911	8-250k	$p \rightarrow CO$	
3G	$\pi^- \rightarrow Ar + 7\% CO_2$; δS only	916	10-250k	$p \rightarrow CO_2$	762
10-80k	^{24}Ng		40k	$^{85}Kr \rightarrow Al_2O_3$ (cryst.)	163
20-160k	$^{85}Kr, ^{125}Xe \rightarrow Al_2O_3, Ta_2O_5, NO_3$		5-160k	$^{24}Na, ^{85}Kr, ^{125}Xe \rightarrow Al_2O_3$	167
20-160k	$^{41}Ar \rightarrow Al_2O_3$		0.5-160k	$^{24}Na, ^{41}Ar, ^{85}Kr, ^{125}Xe \rightarrow Al_2O_3, WO_3$	
20-80K	$^{42}K \rightarrow Ta_2O_5$		40k	$^{135}Xe \rightarrow Ta_2O_5$	182
20-80k	$^{41}Ar \rightarrow NO_3$	819	30k	$^{82}Br, ^{85}Kr, ^{86}Rb \rightarrow Ta_2O_5$	328
0.02-009k	$Cr \rightarrow V_2O_5$	830	40,125k	$^{133}Xe, ^{134}Cs \rightarrow WO_3$	202
40k	^{125}Xe	881	0.5-5.3M	$\alpha \rightarrow CO_2$	790
3-30M	$p \rightarrow Al_2O_3, SiO_2$	904	11.5k	$H \rightarrow SiO_2$ (cryst., amorph.)	800
			6-16k	$H \rightarrow Al_2O_3, Nb_2O_5, Ta_2O_5$	797
			10-80k	^{24}Na	
			20-160k	$^{84}Kr, ^{125}Xe \rightarrow Al_2O_3, Ta_2O_5, WO_3$	
			20-160k	$^{41}Ar \rightarrow Al_2O_3$	
			20-80k	$^{42}K \rightarrow Ta_2O_5$	
			20-80k	$^{41}Ar \rightarrow WO_3$	819
			0.02-0.09k	$Cr \rightarrow V_2O_5$	830
			40k	^{125}Xe	881
			3-30M	$p \rightarrow Al_2O_3, SiO_2$	904

WATER					
Stopping Power			Range		
Energy	Projectiles	Ref.	Energy	Projectiles	Ref.
10-80k	p	99	5.3M	α	3
60-340k	p, d; rel. to air target	40			47
30-600k	p	103			91
4,5M	α	4			51
18-540k	$p \rightarrow D_2O$	130			90
100-800k	Li	605	5.3,7.68M	$\alpha \rightarrow H_2O, D_2O$	30
3-17M	$\alpha \rightarrow La_2Mg_3(NO_3)_{12} \cdot 24H_2O$	661	142,192k	p	33
80-840k	$Li \rightarrow H_2O, D_2O$	813	4-6M	α rel. to air target	89
			5.5M	α	98
			5.3,7.68M	α	29
			1-8.78M	$\alpha \rightarrow H_2O$ (liq. and gas.)	553

TABLE IV

| IONS | | | | | IONS |

PROTONS					^{1}H 1
Stopping Power			**Range**		
Energy	Targets	Ref.	Energy	Targets	Ref.
440k	Li	7	6-18M	Be, Al, Cu, Ag, Au	14
50-600k	Cu, Au, Pb, LiF, CaF$_2$		100-800k	Air	16
75-1400k	Li		880k	Air	20
50-2600k	Be		142,194k	D$_2$ + D$_2$O	33
200-600k	Al, Mn, Ta rel. to Au; Ca, V, Cr, Fe, Co, Ni, Cu, Zn rel. to Mn	8	561k	N$_2$	39
			20-64k	Air, H$_2$	53
4.43M	H$_2$, Air, Kr	26	50-300k	He, N$_2$, Ar, Air	57
1-12M	Air	28	18M	Al	70
400-1050	N$_2$, Ne, Ar, Kr, Xe, Ni, Cu	32	50-250k	He, O$_2$, H$_2$O	92
60-340k	H$_2$, D$_2$, He, H$_2$O rel. to air	40	0.1-2M	Air, Al	95
30-50k	Celluloid	44	1-7M	Fe Cu, Mo, Cd, Sn, Pd, Ta; all rel. to Al	111
0.4-1M	B, Mn, Cu, Ge, Sn, Se, Ag, Sb, Au, Pb	59	10M	Al	114
364,992k	Au	71	16-20.5M	Air	137
0.4-1.35M	Be, Al, Cu, Au, mica	76	1-6M	Al	139
0.5-1.5M	Be	83	1-10k	LiF, NaF, MgF$_2$, CaF$_2$, ZnS	191
0.35-2M	Be, Al, Cu, Ag, mica. Also δS	84	750M	Al, Pb, Cu, U, Emulsion	221
10-80k	H$_2$, He, N$_2$, O$_2$, Ar, Kr, H$_2$O, CO$_2$, CCl$_4$	99	340M	Be, C, Al, Cu, Sn, Pb	209
30-95k	Celluloid	102	100M	Al	227
30-600k	H$_2$, He, O$_2$, Air, N$_2$, Ne, Ar, Kr, Xe, Hydrocarbons	103			592
20.6M	Ni, Cu, Nb, Pd, As, Cd, In, Ta, Pt, Au, Th, all rel. to Al	116	1-25k	Al	193
12M	Ni, Cu, Rh, Pd, Ag, Cd, In, Ta, Pt, Au, Th, all rel. to Al	122	1M	H, Ne, N, O, Ar, Xe	282
50-400k	Be, Al, Cu, Ag, Au	129	4-20k	SiC	288
18-450k	D$_2$O (ice)	130	0.8-1.9M	Si, Ge	312
150-450k	H$_2$, He, Air, Ar	131	57-144M	Cu	379
30-400k	Al, Au	133	5-60k	LiF, ZnS, CsJ	416
4M	Al, Cu, Fe, Mo, Ni, Pt, Ta, Zn all rel. to Air	136	12-40M	H$_2$	445
28.7M	Be, Ti, V, Co, Ni, Cu, Ag, Ta, W, Ir, Au, all rel. to Al	146	0.2-20M	Ar, N$_2$, O$_2$, Air	457
240M	H$_2$, C, O$_2$, N$_2$, Cl$_2$, all rel. to Cu	147	14M	Emulsion	549
19.8M	Be, Ca, Ti, V, Fe, Ni, Cu, Zn, Nb, Mo, Rh, Pd, Ag, Cd, In, Sn, Ta, W, Ir, Pt, Au, Pb, Th, all rel. to Al	149	7.5-52k	H$_2$ → SiO$_2$ (cryst.)	558
			14.7M	Al	561
1.5-4.5M	Be, Al, Ni, Cu, Ag, Au; also δS	151	100k	Si	570
2-15k	Al, Ti, Cu, Ge, Ag, Sn, Au	159	6.5M	Emulsion	581
10-67k	C	166	200-900k	Emulsion	582
2M	Si	171	21.2M	Emulsion	585
40-250k	He, Air, CH$_4$, C$_2$H$_2$, C$_4$H$_4$, C$_3$H$_8$, (CH$_2$)$_3$, C$_3$H$_6$	175	2.47,14.2M	Emulsion	590
37-230k	Cd (vap. and sol.)	176	6.47M	Emulsion	591
2.8M	Si (cryst.)	189	5-50k	Ar, CO$_2$, N$_2$, CH$_4$, C$_2$H$_5$, C$_2$H$_4$, C$_3$H$_8$, C$_4$H$_{10}$	642
992k	C; also δS	190	10-80k	CaF$_2$, RbJ	643
12-25k	C	203	15-50k	SiO$_2$	647
15-55k	Al		1-20k	SiC	682
			0.6-1M	NaF	689
			660M	Cu	700
			0.5-15M	KCl	717
			1.2-13.1M	Emulsion	726
			0.2-16.4M	Emulsion	728
			1-8M	Emulsion	732
			5-16.3M	Emulsion	731
			17-39.5M	Emulsion	733

(CONT'D)

PROTONS (CONT'D)

5-12M	Al	205
		269
65-180k	C	217
350M	H, Li, Be, C, Al, Fe, Cu, Ag, Sn, W, Pb rel. to Al and Cu	218
40-340k	Plastics	219
750M	Al. Pb, U, Emulsion all rel. to Cu	221
635M	H, Be, C, Fe, Cd, W all rel. to Cu	222
4.85M	Si (cryst.)	224
1.5-30k	H_2, D_2	236
30-350k	C, Plastics	232
25-50k	N_2	229
4.25-7.75M	Ge (cryst.)	252
3-11M	Si (cryst.)	257
12-65k	B	266
100-900k	Plastics	273
375k	Si (cryst.)	275
0.1-3.0M	B	278
0.4-6.0M	C, Ca, Au, CaF_2	279
5-12M	Be, Al, Cu, Ag, Pt, Au	280
1M	H, He, N, O, Ar, Xe; δS only	282
14-28k	Au, Au (cryst.)	290
7-35k	Be, Al, Cu, Ag, Au	291
5-24G	Emulsion	292
5-80k	ZnS:Ag	295
3-11M	Si, Ge (cryst.)	305
3-4M	GaSb (cryst.)	307
4-7.6M	Ge, GaAs, Si (cryst.)	308
267,650M	Cu	313
45,730M	Si; δS only	314
10-50k	Au	316
20-95k	Ni	317
4-20k	Cu	318
1.5M	Si (cryst.)	320
20-140k	Al	324
1M	Fluorines	326
5-100k	N_2, Ar	342
400k	Au (cryst.)	343
5-12M	Ca, Sc, Ti, V, Cr, Mn, Fe, Ni, Co, Cu, Zn	358
4-30k	C, Al_2O_3	356
4.5M	Ag, Cd, In, Sn	355
7M	Al	353
0.6-2M	V	310
1.1M	C	369
5-42M	Si; δS only	372
10-20k	CH_4	373
5-13.5M	Al, Ta	374
6.7M	NaCl, KBr (cryst.)	375
29-300M	Si; δS only	385
1.4M	Al (cryst.)	386
0.36-5.49M	Ge	387
25-140k	Cu, Ni, Co, Fe, Mn, Cr	389
4-8M	Si, Ge, Fe, Mo, W, NaCl, MgO, CsI (all cryst.)	391
46M	He-CO_2; δS only	392
		401
		425
0.6-2.4M	Al, Zn, In	398
7-40k	Be, Al, Cu, Ag, Au	399
2-13M	Emulsion. Rel. to air	734
10.8-12M	Emulsion	735
0.2-1.5M	Emulsion	736
0.7-3.2M	Emulsion	738
1-17M	Emulsion	742
12k	Al_2O_3	756
35-120M	Al, Cu, Pb	761
4-250k	H_2	
7-250k	Ar, Air, N_2	
8-250k	CO	
10-250k	CH_4, CO_2	
13-250k	O_2	762
2-21M	Emulsion	763
35k	Ni, Ti	764
1-16k	H, H_2, $H_3 \to$ Cu	765
14-42k	Ni	772
11.5k	SiO_2 (cryst., amorph.), Feldspar	800
10k	Si	801
6-16k	Al_2O_3, Nb_2O_5, Ta_2O_5	797
1.5-60k	Si	817
7.5k	Si(cryst.)	824
2-5.2M	Ni,Ag	903
3-30M	Si,Al,SiO_2,Al_2O_3,$C_3H_5O_2$	904
660M	C,Al,Cu,Sn,Pb	905

(CONT'D)

PROTONS (CONT'D)

4.7,6.7M	NaCl, KCl, KBr (all cryst.; rand. and chann.)	402			
0.4-3.4M	N_2, Air, O_2, Ne, Ar, Kr, CH_4, CO_2	403			
5-12M	Zr, Gd, Ta	404			
100-200k	H_2, He				
50-500k	Air				
100-500k	Ar, Ne				
100-1100k	Kr; all also δS				
1M	H_2, He, Ne, Ar, Kr	406			
4-60k	Al, Cu, Sn, Ag, Au	407			
1.5M	NaCl, KCl, KBr, KI (all cryst., rand. and chann.)	408			
0.5-30k	C, Ti, Al, Cu, Ni, Fe, Ge, Si, Sb, Bi	410			
20-49M	Al, Au; also δS	412			
92k	Au (cryst.)	413			
31.5M	NaI; δS only	414			
1.4-1.8M	BaF_2, CaF_2 (cryst.)	422			
760M	Si; δS only	423			
2-54k	Au (cryst.)	426			
100-500k	H_2, He, Air, Ne, Ar, Kr; also δS	429			
70-90k	C, Al, Ag	430			
20-100k	Au	433			
7.2M	Al, Ni, Cu, Rh, Ag, Pt, Au	435			
7.2M	Al, Ti, Fe, Cu, Mo, Ag, Sn, Ta	443			
17M	Ni; δS only	441			
0.5-1.75M	Ti, Mo, Ta, W	493			
10-30k	Au, Eu_2O_3, Si_2O_3	491			
0.6-2.0M	Fe, Co, Sb; also δS	490			
1-6M	Pb, Ta, Mo, W, Ag, Yb, Ce rel. to 6 MeV p → Pb	486			
0.9-5M	Si (cryst.)	480			
25-250k	Ni, Cu, Ag, Sn, Au	478			
0.7-1.4M	Al, Se, Ag	475			
0.6-10k	Cu, Cr	473			
8-19M	Al; δS only	471			
66-100M	Xe + 10% CH_4; δS only	467			
120k	Si	466			
1.6M	Si (cryst.)	463			
5-24G	Emulsion. Rel. to min.	58			
300-800M	Plast. scint.; also δS	105			
1-4G	Emulsion. Rel. to min.	121			
0.9-5M	Si (cryst. and amorph.)	132			
5-18M	Al, Cu, Ag, Au, Pb, U	499			
200-400k	Si (cryst.)	507			
50-150k	Nb, Ta, Ta_2O_5	508			
4M	CsI (cryst.)	127			
140k	H_2 → Si				
210k	H_3 → Si both rel. to 70 keV H	513			
600M	Ne; δS only	514			
50-160M	Si; also δS	515			
1-30k	O_2, N_2	517			
2.9-6M	Havar	523			
1M	C also 2M H_2 → C	533			
37M	Ar, plast. scint. δS only	535			
50-300k	Si (cryst.) chann. to random ratio	574			
0.7-1.8M	Si (cryst.) chann. to random ratio	577			
200,400k	Si (cryst.)	579			
4.85M	Si (cryst.); Also δS	595			

(CONT'D)

1.01M	C; Also δS	597			
4.7,6.72M	NaCl, KCl, KBr, Si, Ge (all cryst.); also δS	599			
30k	Au (cryst.); Also δS	600			
4.5-6.9M	InAs				
1.9-8M	GaSb				
3.4-6.7M	InSb				
3.9M	AlSb (all cryst.)	601			
2M	Si (cryst.); also δS	602			
6.72M	KBr (cryst.); also δS	604			
0.6-70k	CH₄	612			
13-20G	Plast. scint.	623			
1.5-16G	Ar + 5% CH₄	625			
9G	Ar + 5% CH₄, 60% He + 30% Ar + 10% CH₄, Kr + 5% CH₄	621			
6.3M	W (cryst.)	633			
5-24G	Emulsion. Rel. to min.	645			
6.72M	Si (cryst.); Also δS	646			
0.2-40k	Ag	651			
6.72M	NaCl, KCl, KBr, Si, Ge (all cryst.); Also δS	653			
6.75M	Si, Ge (cryst.); Also δS	654			
1-20k	C	658			
600M	Plastics; δS only	667			
1.15,1.75M	Si (cryst.)	668			
60-300k	C, Au				
75,150k	H₂ → C,Au				
60-100k	H₃ → C, Au all rel. to H⁺	670			
300-400k	BaF₂, CaF₂ (cryst.); Also δS	679			
3M	Si (cryst.)	693			
35-400k	Li	66			
150k	C	247			
30-100k	Ti	442			
0.3-2M	ZnTe	632			
0.6-2.5M	Al	709			
2-20k	Ag	712			
5-17M	Al, Cu	725			
1.6-6.3M	Al, Ag	729			
0.2-1.3M	Li rel. to 900 keV	727			
730M	Si; δS only	730			
40-120k	TiH	737			
2.5-9M	Emulsion	740			
4-16k	Cu, Ag, Au	752			
460k	C; δS only	754			
0.36-4.5M	C	755			
0.5-1.6M	Si (cryst.)	757			
1-16k	H, H₂, H₃ → Cu	765			
1.6-6.8M	Al				
1.6-7.2M	Cu				
1.6-6.3M	Ag				
1.6-5M	Au	779			
0.25-2.5M	Er, Er₂O₃	784			
0.3-2.5M	H₂, N₂, O₂, H₂S	785			
1.5-60k	Si	817			
0.6-1.6M	Al,Cu,Au,Pb,Melinex; δS only	818			
2.2M	Ti,Fe,Ni,Cu,Ag,Au,	821			
40-110k	Au	822			
100-220k	Pt,Au,SiO₂; δS only	844			
0.46-4.79M	C; δS only	853			
5-12M	Al,V,Ni,Mo,Ag,Ta,Au	855			
		866			
25-150G	Ar + 20% Co₂,also δS	868			

(CONT'D)

0.3-1M	H → C				
0.6-2M	H_2 → C	869			
2-60k	Si	871			
1.5M	Al (cryst.)	872			
1.35G	Ge (cryst.,chann.,random)	874			
1.5-6.5M	Ag	885			
6.3M	W (cryst.); also δS	889			
20-260k	Ge,Se,Pd,Ag,Sb,Bi; also δS	896			
80M	Ar + 7% CH_4; δS only	897			
50-150k	H_2,CH_4,CHCH,CH_2CH_2, CH_3CH_3,CH_3CCH,CH_2CCH2, CH_3CHCH_2,$CH_2CH_2CH_2$, $CH_3CH_2CH_3$; also δS	902			
50-150k	same, δS only	913			
1.4-4.6M	Au	908			
2-13k	Cu; also δS	910			
6.3M	W (cryst., chann. to random ratio)	919			

DEUTERONS					^{2}H 1
	Stopping Power			Range	
Energy	Targets	Ref.	Energy	Targets	Ref.
8.86M	H_2, He, N_2, O_2, Ne, Ar, Kr, Xe	26	5-27k	Al, Cu, Ag	140
60-340k	H_2, D_2, He, Air;, H_2O	40	1-25k	Al	193
9M	Si, Ni, Cu. Ge, Zr, Rh, Ag, Sn, Air all rel. to Al	67	2-9k	Al	206
150-450k	H_2, He, Air, Ar	131	4-20k	SiC	288
30-650k	Al, Au	133	5-80k	Au, Al	337
1.5-4.5M	Al, Ni, Cu, Ag, Au; also δS	151	7.5-52k	D_2 → SiO_2 (cryst.)	558
2-15k	Al, Ti, Cu, Ge, Ag, Sn, Au	159	28M	LiF	571
12-50k	Al	203	0.6-1.5M	Emulsion	582
5-12M	Al	205	5,5.5M	Emulsion	585
		269	40k	Al, Au	
13k	Au (cryst.)	225	80k	D_2 → Al, Au	
1.5-30k	H_2, D_2	236	120k	D_3 → Al, Au rel. to D$^+$	594
0.4-1M	Ag	250	2-20k	SiC	682
7.63M	Ge (cryst.)	252	1,4k	Zr	832
12-60k	B	266	100-400k	Ni	847
100-900k	Plastics	273	400k	Ni (cryst. and polycryst.)	917
15k	Au (cryst. and polycryst.)	284			
3-4M	GaSb (cryst.)	307			
4-7.6M	Ge, GaAs, Si (all cryst.)	308			
27M	Al	329			
377M	Cu	313			
16-30k	Au (cryst.)	340			
10-50k	N_2				
10-80k	Ar	342			
5-12M	Ca, Sc, Ti, V. Cr, Mn, Fe, Co, Ni, Cu, Zn	358			
14M	Al	353			
5-13.5M	Al, Ta	374			
7-40k	Be, Al, Cu, Ag, Au	399			
5-12M	Zr, Gd, Ta	404			

(CONT'D)

DEUTERONS (CONT'D)

2-54k	Au (cryst.)	426
2.5-7M	Havar	427
20-28k	Au (cryst.)	428
14.4M	Al, Ni, Cu, Rh, Ag, Pt, Au	435
30-100k	Ti	442
0.9-5M	Si (cryst.)	480
0.9-5M	Si (amorph. and cryst.)	132
28M	LiF. Dist. of energy along path	134
5-18M	Al, Cu, Ag, Pu, Pb, U	499
0.8-3.9M	Havar	523
28M	LiF	571
2.5-8M	GaSb (cryst.)	601
2.4,3.5M	Havar, μ-metal, permalloy	619
0.2-40k	Ag	651
1-20k	C	658
5-17M	Al, Cu	725
1.6-9M	Al, Ag	729
1.6-9M	Al,Ag	
2.3-8.5M	Cu	
1.7-8.4M	Au	779
100-200k	Au,Pt,SiO$_2$; δS only	843
500k	Nb	850
2.6-8.5M	Ag	885
16-27M	LiF (cryst.); also δS	892
3-6.8M	Ag,Au	894
2.4-6.8M	Al	
3-5.8M	Cu	
2.8-6.5M	Ag	908
3.4-8.5M	Au	908

TRITONS					^{3}H 1
Stopping Power			Range		
Energy	Targets	Ref.	Energy	Targets	Ref.
0.2-2.73M	N$_2$, Al, Ar, Ni, Kr, Xe, Air	142	2.75M	Air	37
10-30k	H$_2$	99	0.2-2.73M	N$_2$, Al, Ar, Ni, Kr, Xe, Air	142
			30.6M	He	
			8.43M	^{4}H $\rightarrow$ He	418
			2.5,5M	Emulsion	585

HELIUM					He 2
	Stopping Power			Range	
Energy	Targets	Ref.	Energy	Targets	Ref.
4-5M	H_2O rel. to Air	4	5.3M	H_2O, C_6H_6, C_2H_4OH, pyridine, n-heptane, iso-octane, 2-2-4-trimethyl-pentane	3
5.3,6.1M	H_2^+, He, O_2, Ne, Ar, Kr, Xe rel. to air	61	480-615k	Air	9
5.3M	Al, mica, nylon, polystyrene rel. to air	63	6.1M	He. Ne, O_2, Ar, Kr, Xe	10
5.3M	H_2, N_2, O_2, N_2O, H_2S, C_3H_8, C_3H_6, C_4H_{10}, C_4H_8, CCl_4, C_2H_5Cl, CCl_2F_2, C_2H_5I	65	0.25-3.5M	Ar	11
28,37M	Cu, Ag, Ta, Bi, Th rel. to Al	77	100-500k	Air	16
4.2-7.7M	H_2, He, Ne, Ar, Air	85	5.3M	H_2, He, N_2, Ar, Ar+1.5% CO_2, Ar+10% CO_2, CH_4, Ar+10% CH_4, Kr, Kr+1% CO_2	23
5-8M	Al, Cu, Au, Air, mica	88	7.7M	CH_3Br, CH_3I, C_2H_5Cl, CCl_4, C_2H_5O, C_2H_5, H_2, Al, Cu, Ag, Sn, Pt, Au. All rel. to air	24
6M	CH_2, plastics, Al rel. to air	96	3-5.3M	Mica rel. to air	25
3-7M	He, Ne, Ar, Kr, Xe rel. to air	93	5.3,7.7M	H_2O. Liq. and sol.	29 30
1-4M	Mica	106	4.7M	Air	31
5.3-7.7M	Li, Al, Fe, Cu, Zn, Mo, Pd, Ag, Cd, Sn, Pt, Au, Pb, Mica, AuAg alloys, AgCu alloys	110	5.3M	Air, Ar	35
7.7M	Au, Sn, Pb, Al, H_2, paper, collodium rel. to air	117	1.5-4.5M	H_2, He; rel. to air	36
150-450k	H_2, He, Air, Ar	131	1.47M	Air	37
30-1400k	Al, Au	133	20-250k	He, N_2, Ar, Air	48
5.3M	$(CH_2)_n$. Sol. rel. to gas	135	0.7-4.45M	Si, Ge, InSb, Al, Cu, Ag, Au	56
8-40M	H_2, N_2, Ar	148	50-300k	He, N_2, Ne, Air	57
2-15k	Al, Ti, Cu, Ge, Ag, Sn, Au	159	5.3M	CF_4, SF_6, rel. to air	60
10-80k	C	166	1-3M	Air	62
40-250k	C_2H_4, C_3H_8	175	5.3-7.7M	H_2, Ne, O_2, Air	64
40-460k	H_2	177	4.4M	Air rel. to 4.2M	81
148-920k	Cd (gas. and liq.)	176	5.3M	Air, CO_2, N_2, O_2	82
15-65k	Al	203	4-8M	Al, Cu, Ag, Sn, Pt, Au, Mica rel. to air	87
40-200k	CH_4, C_2H_2, C_2H_4, C_3H_8, $C_3H_6(CH_3)_2$	215	4-6M	H_2O rel. to air	89
65-180k	C, Al, Cr	217	5.3M	Air, H_2, H_2O, CH_4, N_2, O_2, CO, CO_2, SO, SO_2, CH_3Br	90
40-340k	Plastics	219	5.3M	CS_2, H_2O, C_2H_5OH, ether, C_6H_6, anilin, glycerine, chloroform	91
8-40M	O, Ni, Emulsion	220	100-360k	He, O_2, H_2O	92
30-350k	C, Plastics	232	5.5M	H_2O, CO, CO_2, CH_3Br, CH_3I, Cl_2, HCl, NH_3; rel to air	97
25-50k	N_2	229	5.5M	Air, CO_2, H_2O, C_2H_5OH, C_6H_6, C_2H_5N	98
1-8M	H_2, C, Hydrocarbons	244	0.5-4M	SF_6	107
600-950k	Al		5.3M	Air, N_2, O_2, NO, N_2O, NO_2, CO, CO_2, CH_4, C_2H_4, Ne, Ar, Kr, N_2+O_2, N_2+2O_2, $2N_2+O_2$	112
400-950k	Ni, Ag		5.3,5.5M	H_2, He, O_2, air	118
350-950k	Au	248	7.7M	H_2, He, Li, O_2, Mg, Al, Ca, Fe, Ni, An, Zn, Ag, Cd, Sn, Pt, Cu, Tl, Pb	123
0.3-1.3M	C		5.3M	Air rel. to 4.2M	138
0.3-2M	Al, Ni, Ag, Au	250	4-40M	Al	141
15-70k	B	266	5.3M	C	167
3M	Au (cryst.)	267	1-10k	LiF	192
28M	Au, Sn, Mylar	268	1-10k	LiF, NaF, MgF_2, CaF_2, ZnS	191
2-9M	Al, Ni, Ag, Au; also δS	274			
0.2-5.5M	Hydrocarbon	277			
0.2-5.3M	B	281			
15k	Au, Au (cryst.)	284			
14-28k	Au, Au (cryst.)	290			
10-100k	ZnS:Ag	295			
3-4M	GaSb (cryst.)	307			
765M	Cu	313			
910M	Si; δS only	314			

(CONT'D)

HELIUM (CONT'D)

20-95k	Ni	317	1-25k	Al	193
20-140k	Al	324	8-40M	O, Ni, Emulsion	220
10-100k	ZnO:Zn rel. to H$^+$	325	8-40M	H$_2$, N$_2$, Ar	208
50-54M	Al, Cu, Ag, Tb, Tm, Au; also δS	329	0.5-5.6M	Air, N$_2$, O$_2$, CH$_4$, CO$_2$	251
10-90k	N$_2$, Ar	342	4-20k	SiC	288
800k	Au (cryst.)	343	54.4M	Cu, Ag, Tb, Tm, Au	329
4-30k	C, Al$_2$O$_3$	356	0.14-2.2M	H$_2$, He, CH$_4$, Benzene, Air, Ar	362
0.14-2.8M	H$_2$, He, CH$_4$, Benzene, Air, Al, Ar, Ni, Ag, Au	362	5-60k	LiF, ZnS, ZnJ	416
0.5-3.5M	Cu, Au, Cu-Au alloys	368	0.2-2M	Ar, N$_2$, O$_2$, Air	457
10-120k	CH$_4$	373	1.47,1.78M	Air, Emulsion	126
8-20M	^{3}He, ^{4}He → Al, Ta	374	0.5-2M	Air	50
1-9M	Ar	381	5.3M	Kr, Xe	511
0.4-2M	Be, C, Mg, Al, Ti, V, Cr, Mn, Fe, Co, Ni, Cu, Gd, Pd, Ag, In, Sn	382	5.3,7.7M	Air	512
3M	Au (cryst.)	384	0.5-4M	Emulsion	534
1-3.5M	He, Air, Ar, Kr, Xe; also δS	388	87,118,146M	Emulsion	547
40-120k	Cu, Ni, Co, Fe, Mn, Cr	389	2-40M	Emulsion	551
100-600k	Hc; also δS			H$_2$O (Liq. and vap.)	553
100-500k	Air; δS only		1-8.78M	Emulsion	556
200-400k	Ne; δS only	406	4.2,4.4M	SiO$_2$ (cryst.)	558
1-9M	Al, Ni, Se		7.5-52k	21 hydrocarbons	560
1-5.3M	Ag, Au	411	5.3M	C$_2$H$_5$OH, CCl$_4$ (both liq. and vap.)	562
80M	Al, Au; also δS	412	1-8.9M	Au	565
5.5-8.8M	Al$_2$O$_3$	424	9.6-14.4M	He, Ne, Ar, Kr, Xe, H$_2$, N$_2$, O$_2$, NH$_3$, CO, CO$_2$, NO, N$_2$O, CH$_4$, C$_2$H$_6$, C$_3$H$_8$, C$_4$H$_{10}$	567
2-54k	Au (cryst.)	426	5.3M		
0.5-2M	H$_2$, D$_2$, N$_2$, Ar	431	8.78M	Air, He, Ne, Ar, Kr, H$_2$, N$_2$, O$_2$, CO, CO$_2$, CH$_4$, C$_2$H$_6$, C$_3$H$_8$, C$_4$H$_{10}$	568
20-100k	Au	433	9.02-25.2M	Emulsion	569
3-15M	Ni, Ge, Y, Ag, Au	434	56M	LiF	571
28M	Be, Al, Cu, Mo, Ta, Au	436	1.2-2.4M	Emulsion	582
0.1-5.3M	N$_2$, O$_2$, CH$_4$, CO$_2$	437	5,10M	^{3}He → Emulsion	
5M	N$_2$, O$_2$, CO, CO$_2$, NH$_3$, hydro-carbons	438	5M	Emulsion	585
0.3-2M	H$_2$, O$_2$, N$_2$, NH$_3$, N$_2$O, CO, CO$_2$, CH$_4$, C$_2$H$_2$, C$_2$H$_4$, C$_2$H$_6$, C$_3$H$_6$, (CH$_3$)$_2$	439	50-150k	Cu	
		440	50k	Ti, V, Nb	607
0.3-1.7M	Si, SiO$_2$	492	10-80k	CaF$_2$, RbJ	643
1-25M	W	486	2-20k	SiC	682
6-30M	^{3}He → Si		0.4-2M	ZnTe	632
8-40M	^{4}He → Si	484	1.5-15k	^{3}He → Nb	714
3M	Au, Ag (cryst.); also δS	483	2.1-13M	Emulsion	726
0.1-18M	Si (cryst. and amorph.)	482	1.1-18.9M	Emulsion	728
0.9-5M	Si (cryst.)	480	1-8M	Emulsion	732
0.42-2.75M	Si (cryst. and amorph.)	474	8-19M	Emulsion	731
5.2M	He, Ne, Ar, Kr, CH$_4$, CO$_2$, N$_2$, Air; δS only	472	5-9M	Emulsion rel. to air	734
1-9M	ThO$_2$, UO$_2$	470	1-5.3M	Emulsion	738
1-5M	U, UC, (U,Pu)C; also δS	469	5.3M	H$_2$, O$_2$, H$_2$O	51
0.3-2M	CF$_4$, C$_2$F$_6$, C$_3$F$_8$, CCl$_4$, CClF$_3$, CCl$_2$F$_2$, CBrF$_3$, C$_2$H$_3$Br, C$_2$H$_5$Br, C$_2$H$_5$I	468	1.53-2.97M	Air	54
0.4-1.9M	Au	465	18-38M	Al	768
1-8M	Ar; also δS	464	5.3M	NaCl, KCl, KBr, KJ (all cryst.)	771
5-7M	AuAg and AgAu alloys	110	4.5M	Si, Zr, Au, U	773
2M	Al, Si, V, Fe, Co, Ni, Cu, In, Ge, Mo, Sb, Te, Gd, Hf, Ta, W, Ir, Pt, Au, Pb	124	20-250k	H$_2$, Ar, Air	762
0.9-5M	Si (cryst., amorph.)	132	3-16M	Emulsion	788
56M	LiF	134	0.5-5.3M	H$_2$, He, CH$_4$, CO$_2$	790
		571	4k	Nb (cryst., chann. and random)	805
			15k	^{3}He→Nb (cryst., chann.)	
			1.5k	^{3}He→Nb (cryst., random)	849
			15k	^{3}He→Nb (cryst.)	850
			54-158k	Cu	856
					857
			2-8M	N$_2$, O$_2$, Ar, Air	901

(CONT'D)

HELIUM (CONT'D)

0.3-2M	Si, Ge (amorph.)	497			
0.3-2M	Se, Y, Zr, Nb, Mo, Sb, Te, La, Dy, Ta, W, Au	500			
2M	Ti, V, Ni, Mo, Ta	501			
0.25-2.5M	Er	502			
1-2M	Ag rel. to Au, Cu rel. to Ag, Cu rel. to Au, Al rel. to Au, Si rel. to Al	503			
0.3-2M	CO, CO_2, C_2H_3Br, C_2H_5Br, $CBrF_3$, $C_2Br_2F_4$, $(CH_3)_2O$, $C_2H_2F_2$, hydrocarbons	504			
2M	NbN, NbC	505			
0.5-2.25M	AuAg, AuCu, AuAl alloys	506			
1-2M	Fe_2O_3, Fe_3O_4, Al_2O_3	509			
8.8M	Pt; δS only	516			
1M	Si, Ge (cryst.)	517			
0.6-2M	Si (cryst.)	521			
1.5-8M	Al, Ni, Pt, Au	526			
0.4-1.9M	Organic compounds (sol., liq., gas)	548			
3.72,4.33M	Bi, Pb, W	550			
6.05,6.89M	Al, Ni, Ag, Au (all rel. to air)	566			
5.3M	Au	567			
	He, Ne, Ar, Kr, Xe, H_2, N_2, O_2, NH_3, CO, CO_2, NO, N_2O, CH_4, C_2H_6, C_3H_8, C_4H_{10} (all rel. to air)				
8.78M	He, Ne, Ar, Kr, H_2, N_2, O_2, CO, CO_2, CH_4, C_2H_6, C_3H_8, C_4H_{10} (all rel. to air)	568			
5-9M	Al, Ni, Au	580			
5.3M	Al, mica; δS only	589			
0.3-5M	Al, Ag, Au, Zapon, Paraffine	598			
5.4-13.9M	$^3He \rightarrow$ GaSb (cryst.)	601			
0.6-70k	CH_4	612			
5.3M	Al, Cu, Ag, Au rel. to air	618			
5.4M	Havar, μ-metal, mylar, teflon	619			
7.7M	Si; δS only	628			
0.96-5.3M	CdS	629			
0.3-2M	ZnTe	632			
0.5-1.75M	Ti, Mo, Ag, Ta, W	634			
6-16k	Cu	637			
1-9M	Ar	657			
1-20k	C	658			
3-17M	$La_2Mg_3(NO_3)_{12} \cdot 24H_2O$	661			
6.8M	Al, Cu, Ag, Au, Melinex; δS only	665			
3-5.3M	Air, CO_2, He	666			
2M	Au (cryst.); also δS	669			
5.5M	Al; δS only	674			
0.3-2M	He, Ne, Ar, Kr, Xe	675			
300k	Al	677			
6.62M	Ne, Kr, Ar; also δS	680			
3-5.5M	CH_4, CO_2, CF_4, CF_2A_2, SF_6	687			
1-5.5M	TEP (Tissue equivalent plastic)	691			
1-8M	CCl_4, hydrocarbons (liq. and sol.)	694			
1-2M	Al, Ni, Au δS only	704			
1-5M	C, H_2, C_2H_6, C_4H_{10}, C_2H_4, C_3H_6, C_4H_8, C_2H_2, C_3H_4, C_4H_6, CH_4, C_6H_6, ethylene, polyethylene, propylene, polypropylene	233			

(CONT'D)

100-500k	Air, He, Ne; also δS				
100-300k	H_2, O_2; also δS	421			
5-8.7M	Si; δS only	711			
2-20k	Ag	712			
2.2-18.6M	$^3He \rightarrow Al$				
3.2-17.2M	$^4He \rightarrow Al$				
2.0-13.5M	$^3He \rightarrow Ag$				
3.2-19.2M	$^4He \rightarrow Ag$	729			
910M	Si; δS only	730			
4-16k	Cu, Ag, Au	752			
0.25-3.0M	CdS	778			
2.2-18.6M	$^3He \rightarrow$				
3.2-17.2M	$\alpha \rightarrow Al$				
5-18M	$\alpha \rightarrow Cu$				
3.0-13.5M	$^3He \rightarrow$				
3.2-19.2M	$\alpha \rightarrow Ag$				
4.1-15.5M	$\alpha \rightarrow Au$	779			
28.8M	Al, Ti, Fe, Ni, Cu, Mo, Ag, Ta, Au	781			
8.78M	Al, Cu, Ag, Ta	783			
0.25-2.5M	Er, Er_2O_3	784			
0.3-2.5M	H_2, N_2, O_2, H_2S	785			
0.2-2.0M	Au, Pt, Ta_2O_5, SiO_2	786			
0.3-5.3M	N_2, Air, Kr, CO_2, CH_4	793			
1.3,6.6M	$\alpha \rightarrow Si$ (cryst.); Also δS	799			
0.3-2M	Al	816			
0.3-2M	Ag,Au	820			
40-110k	Au	822			
1-2M	MgO, Al_2O_3, SiO_2, Fe_2O_3, Fe_3O_4	824			
2.54-4.93M	H_2				
2.09-5.29M	Ne				
0.96-5.13M	He				
0.5-5.11M	C_2H_6				
0.59-4.44M	C_3H_8				
0.82-422M	C_4H_{10}	828			
250k	Au, Pt, SiO_2; δS only	843			
0.3-1.7M	Al_2O_3; δS only	846			
2M	Fe_2O_3, Fe_3O_4, MgO, Al_2O_3, SiO_2, Si_3N_4	851			
0.5-2M	Al, Cu; also δS	854			
6.11M	Al, Ag, Au; δS only	862			
0.5-2M	Ti, Cr, Co, Cu, Ag; δS only	865			
2-60k	Si	871			
1.3M	Al, Cu, Ag, Au; δS only	878			
0.3-2M	C	879			
2M	Fe_2O_3, Fe_3O_4, MgO, Al_2O_3, SiO_2, Si_3N_4 all rel. to metal (semicond.)	880			
0.9-77M	Si; δS only	882			
1-2M	Si; δS only	890			
6.05,8.78M	Anthacene; also δS	893			
5-13M	Ag, Au	894			
20-260k	Ge, Se, Pd, Ag, Sb, Bi; also δS	896			
0.3-2M	$C_2H_2F_2$, $C_2H_4F_2$, C_3H_8, C_4H_6, C_4H_{10}, C_3H_4	899			
0.2-8.5M	Polycarbonate	906			
0.5-7.5M	Polycarbonate	907			
4.6-13.6M	Al				
5-11M	Cu				
5-13M	Ag				
6-12M	Au	908			

(CONT'D)

6-12k	Cu; δS only	910			
8.78M	C, Au, Ni, Al, SiO₂, Ru; δS only	911			
0.5-5.5M	(73.7% H₂ + 25% Co₂ + 1.3% N₂), (63.4% CH₄ + 33.4% CO₂ + 3% N₂), Air, Melinex, Mylar; also δS	912			

LITHIUM					Li 3
	Stopping Power			Range	
Energy	Targets	Ref.	Energy	Targets	Ref.
100-450k	H₂, He, Ar, Air	2	40-450k	^{8}Li → H₂, D₂, He	210
2.74M	Al, Cu, Au	42	4-22M	^{8}Li → Emulsion	207
750-850k	Al, Au	133	2-12k	Cu, Ag	359
20-70k	^{6}Li → C		5-15k	Au	361
15-70k	^{7}Li → C	166	0.2-4.2M	H₂, He, CH₄, Benzene, Air, Ar	362
20-145k	C	194	0.5-5M	H₂, Air, Emulsion	371
20-70k	^{7}Li → Al	203	0.84,1.02M	^{7}Li → Air, Emulsion	126
15-70k	^{7}Li → B	266	220k	Ag, Nb	578
3-20k	Cu	346	0.75-1.4M	Emulsion	582
0.2-4.2M	H₂, He, Al, Ar, Ni, Cu, Ag, Air, Benzene	362	2.4-6k	ZnS	584
0.5-5M	^{7}Li → H₂, Air, Emulsion	371	20-6-22.4M	^{8}Li → Emulsion	586
0.6-3.8M	^{6}Li → H₂, He, CH₄, N₂, CO₂	370	0.75-2M	^{6}Li → Emulsion	587
10-120k	CH₄	373	9.02-25.2M	^{7}Li → Emulsion	588
30-100k	C, Ar, Ti, Ni, Cu	395	220k	Nb	608
0.06-3.0M	Ar	405	5-25k	Ni	676
100-500k	He; also δS		2-20k	SiC	682
200-300k	Air; δS only	406	0.85-1.8M	^{7}Li → Emulsion	738
10-20M	Ar	409	0.5-7M	^{8}Li → Emulsion	739
3-10k	C; dep. on scatt. angle	420	1.6-5.3M	^{7}Li → Emulsion	741
2-54k	Au (cryst.)	426	2-11k	SiC (cryst.)	767
50k	C	488	0.1-5.0M	Air	770
10-46k	C	479	10-45k	Diamond	809
100-800k	B₄C, B, H₂O, H₃BO₃, MoB, WB	605			
2-120k	^{6}Li, ^{7}Li → CH₄	612			
10-100k	Si, V, Cr, Fe, Ge, Se	655			
1-20k	C	659			
5-16k	Ag	683			
100-500k	He, Ne, Air; also δS	421			
70k	He, Ar, N₂, O₂	776			
14-20M	^{6}Li →				
8-23M	^{7}Li → Al				
10-17M	^{7}Li → Cu				
8.5-21M	^{7}Li → Ag				
13-15M	^{6}Li →				
10-21M	^{7}Li → Au	779			
175k	CrB, CrB₂, NbB, NbB₂, MoB, MoB₂, TaB, TaB₂, WB, W₂B₅	812			

(CONT'D)

LITHIUM (CONT'D)

Energy	Targets	Ref.			
80–840k	Be, B, Al, Ti, Cu, Ta, AlB_2, AlB_{12}, B_4C, B_2O_3, BPO_4, B_4Si, CaB_6, CeB_6, CrB, CrB_2, Cr_5B_3, H_2O, D_2O, HBO_2, H_3BO_3, HfB_2, KBF_4, KBH_4, LaB_6, $LiBH_4$, MoB, MoB_2, $Na_2B_4O_7$, $Na_2B_4O_7 \cdot 10H_2O$ (borax) $NaBH_4$, NbB, NbB_2, NH_4BF_4, TaB, TaB_2, TiB_2, VB_2, WB, W_2B_5, YB_6, ZrB_2	813			
80–840k	CrB_2	814			
70,110k	B				
100k	Al, Ti, Cu, Ta, C, TaB, TaB_2, CrB, CrB_2, NbB, NbB_2, MoB, MoB_2, WB, W_2B_5, LiOH, HCl, KNO_3, H_2SO_4, KF, KCl, KBr, KJ, LiCl, $LiNO_3$, $ZnCl_2$, $Zn(NO_3)_2$, $LaCl_3$, $La(NO_3)_3$, $CeCl_3$, $Ce(NO_3)_3$ (all elect. dep. on conc.)	815			
8.5–21M	$^7Li \rightarrow$ Ag, Au	894			
13–15M	$^6Li \rightarrow$ Au				
14–20M	$^6Li \rightarrow$ Al				
10–21M	$^7Li \rightarrow$ Au				
8.5–21M	$^7Li \rightarrow$ Ag				
10–17M	$^7Li \rightarrow$ Cu				
8–23M	$^7Li \rightarrow$ Al	908			

BERYLLIUM					Be 4
Stopping Power			Range		
Energy	Targets	Ref.	Energy	Targets	Ref.
12–130k	C	166	2–20M	Au, Al	150
0.25–5.6M	H_2, He, CH_4, Benzene, Air;, Al, Ar, Ni, Ag, Au	362	0.3–5.6M	H_2, He, CH_4, benzene, air, Ar	362
10–120k	CH_4	373	3.1,3.7M	Air	554
4–15M	Ar	409			
200–500k	He, Ne, Air also δS	421			
3–120k	CH_4	612			

BORON					B 5
	Stopping Power			Range	
Energy	Targets	Ref.	Energy	Targets	Ref.
12-140k	$^{11}B \rightarrow C$	166	10-110M	Al	141
15-150k	$^{11}B \rightarrow Al$	203	20-100M	$^{10}B \rightarrow$	
20-100M	$^{10}B \rightarrow$		22-110M	$^{11}B \rightarrow O, Ni, Emulsion$	220
22-110M	$^{11}B \rightarrow O, Ni, Emulsion$	220	8-16M	$^{8}B \rightarrow Emulsion$	207
5-20M	Ag, Ni, Au	223	10-70k	Si (cryst.)	333
		264	30-100k	SiO_2	349
15-140k	$^{11}B \rightarrow B$	266	0.5-7M	$^{11}B \rightarrow Air$	78
100-500k	Si (cryst.)	339	0.35-7M	H_2, He, CH_4, Benzene, Air, Ar	362
20-160k	$^{11}B \rightarrow N_2$		200-400k	Si (cryst.)	376
25-160k	$^{11}B \rightarrow Ar$	342	0.15-1.8M	Si (cryst.)	380
0.35-7.0M	H_2, He, CH_4, Benzene, Air, Al,		15k	Si	459
	Ar, Ni, Ag, Au	362	30-300k	$^{11}B \rightarrow Si$, Si (cryst.)	456
10-120k	CH_4	373	50-200k	Si	454
0.06-3.0M	Ar	405	40-500k	Si	453
200-500k	$^{11}B \rightarrow He$; also δS		10,40k	Si	1
200-300k	$^{11}B \rightarrow Air, Ne$; δS only	406	20k	$^{11}B \rightarrow Si$	5
6-20M	Ar	409	70k	Si	6
50k	C	488	30-75k	Si	17
15-46k	C	479	100k	Si (amorph.)	18
3-120k	$^{11}B \rightarrow CH_4$	612	20-22k	$^{11}B \rightarrow Si$ (amorph.)	19
200-500k	Air, He, Ne; also δS	421	50-300k	Si	128
2M	$^{11}B \rightarrow Si$	817	40-300k	$^{11}B \rightarrow SiO_2$	518
2-60k	Si	871	30-200k	Si (polycryst.)	
			70-800k	Si (amorph.)	520
			60-100k	SiO_2	
			20,40k	Si_3N_4	
			60k	Si (amorph.)	537
			50-150k	$^{10}B \rightarrow Si$ (cryst. and amorph.)	538
			100k	Si	541
			30-70k	Si	542
			10-250k	Si	543
			5-100M	$^{10}B \rightarrow$	
			5.5-110M	$^{11}B \rightarrow Emulsion$	551
			70-280k	Si	552
			2.2-6.3M	$^{11}B \rightarrow Emulsion$	564
			70k	Si	614
			100-400k	Si (cryst.; chann. and rand.)	616
			100-300k	Si	617
			30-200k	Si (cryst.)	620
			25-125k	Si	649
			1-2.5M	Si	650
					698
			20-60k	Si	656
			100k	Si (cryst.; chann. and rand.)	671
			100k	Si	690
					692
			40,100k	Si	695
			30-100k	Si (cryst.)	721
			60-200k	^{10}B, $^{11}B \rightarrow Si$	746
			5-150k	$^{11}B \rightarrow SiO_2$	747
			50-150k	$^{10}B \rightarrow Si$	748
					750
			60,100k	Si	789
			30-800k	Si (amorph., polycryst.)	802
			30k	Si	806
			40-250k	Diamond	811

(CONT'D)

BORON (CONT'D)

				34k	Si	825
				6k	Si	831
				80,150k	$^{11}B \rightarrow S$; also $\delta R_\perp$	835
				75-250k	$\delta R_\perp$ only	836
				120k	Si	839
				50-200k	^{10}B, $^{11}B \rightarrow$ Si	841
				20-80k	Si	859
				5-30k	Si	914

CARBON C 6

Stopping Power			Range		
Energy	Targets	Ref.	Energy	Targets	Ref.
20-120M	H_2, He, N_2, Ar	148	0.5-5M	$^{13}C \rightarrow$ Air	78
12-140k	C^2	166	12-120M	Al	141
15-150k	C	203	24-120M	O, Ni, Emulsion	220
24-120M	O, Ni, Emulsion	220	12-120M	Emulsion	214
25-50k	N_2	229	50-110M	^{12}C, $^{14}C \rightarrow$ Al, Cu, Au	213
82-380k	C^2	247	24-120M	H_2, He, CH_4, N_2, Ar	208
0.36-3.2M	C, Al, Ni, Ag, Au	249	0.55-1.64M	$^{11}C \rightarrow$ Al	234
15-140k	B	266	0.4-1.9M	Be	350
100-500k	Si (cryst.)	339	0.42-8.4M	H_2, He, CH_4, Benzene, Air, Ar	362
100-500k	Air	341	6-120M	Emulsion	551
20-180k	N_2		2.2-5.0M	$^{13}C \rightarrow$ Emulsion	564
25-160k	Ar	342	2-12M	$^{13}C \rightarrow$ Ni, Ta	688
0.4-1.9M	Be	350	130k	$^{11}C \rightarrow$ Polystyrene	723
0.42-8.4M	H_2, He, CH_4, Benzene, Air, Al, Ar, Ni, Ag, Au	362	40k	Emulsion	738
12-120k	CH_4	373	40-112M	Emulsion	787
50k	H_2, D_2, Ne; also δS	396	40k	Au	881
200-400k	He: also δS				
100-400k	Air, Ne; δS only	406			
8-25M	Ar	409			
200-500k	Air, He, Ne; also δS	421			
22-46k	C	479			
0.5-2.0M	Cr, Fe, Co, Ni, Cu, Zn	481			
24-120M	Si	485			
50k	C	488			
0.4-1.5M	C	15			
3-120k	CH_4	612			
2-20k	Ag	712			
12-120M	Mylar, Polythene	722			
0.3-3M	Si; also δS, $\eta(\varepsilon)$	780			
300k	Nb, Mo, Ta, Ag	815			
0.3-1.7M	Al_2O_3; δS only	846			
2-60k	Si	871			
8-40M	Si; δS only	882			

NITROGEN					N 7
Stopping Power			**Range**		
Energy	Targets	Ref.	Energy	Targets	Ref.
150-450k	H_2, He, Air, Ar	131	20-250k	N_2, He, Ar, Air	48
15-140k	C	166	8-29M	Ni	104
20-130k	Al	203	50-300k	He, N_2, Ar, Air	57
28-140M	O, Ni, Emulsion	220	14-140M	Al	141
25-50k	N_2	229	0.4-6.4M	Au	152
73-418k	C	247	50-500k	Be, B, C, Al	164
0.46-1.5M	Al		28-140M	O, Ni, Emulsion	220
0.6-1.8M	Ni, Au	248	14-140M	Emulsion	214
0.4-3.8M	Ag, Au		4-28M	Emulsion	212
0.4-5.6M	Ni		4-28M	Al	211
0.4-5.2M	Al		0.4-2.5M	Ni, Ag	235
0.36-2.6M	C	250	3-120M	Emulsion	253
7-28M	Ni, Ag, Au	264	10-70k	Si (cryst.)	333
15-140k	B	266	0.5-2.0M	Be	350
12-80k	ZnS:Ag	295	0.5-10M	H_2, He, CH_4, Benzene, Air, Ar	362
10-100k	ZnO:Zn rel. to H^+	325	1-12M	Al, Ni, Ag, Au	366
100-500k	Si (cryst.)	339	0.3-2M	Air	451
100-500k	Air	341	50-270M	Polyethylene	
20-200k	N_2		44-270M	Polymethymetacrylat	
25-160k	Ar	342	270M	Al, Cu, Pb	448
0.5-2M	Be	350	7-140M	Emulsion	551
10-120k	CH_4	373	4.9-9.5M	Air	554
0.53-10M	H_2, He, CH_4, Benzene, Air, Al, Ar, Ag, Au	362	· 5-30k	CH_4	647
300-900k	Au (cryst.)	390	1-1.6M	Si	650
0.1-7M	Ar	405	50k	GaP, Si	743
200-500k	He; also δS		46-82k	Ge	769
200-450k	Air, Ne; δS only	406	800k	$22 \leq Z_2 \leq 32$	798
10-16M	Ar	409	800k	$22 \leq Z_2 \leq 32$; $Z_2 = 40, 41, 42$	808
1.9-10M	Si	411			
200-500k	He, Ne, Air; also δS	421			
50k	C	488			
15-46k	C	479			
0.5-2.0M	Cr, Fe, Co, Ni, Cu, Zn	481			
28-140M	Si	485			
1-10M	Sb	555			
3-120k	CH_4	612			
1-20k	C	658			
5.7M	Si (cryst.)	668			
0.3-2.0M	Si; δS, η(ε)	710			
2-20k	Ag	712			
100,200k	C	719			
0.3-3M	Si; also δS, η(ε)	780			
1.0-7.4M	T, Ge, Ni, Ag, Au, W	782			
10M	N_2; also δS	792			
4.4M	Si (cryst.); also δS	799			
800k	$22 \leq Z_2 \leq 32$	798			
40-110k	Au	822			
0.3-1.7M	Al_2O_3; δS only	846			
800k	Z_2 = 22-32, 40-42	848			
2-60k	Si	871			
4.6M	Al, Cu, Ag, Au; δS only	878			
5-48M	^{14}N				
2-100M	$^{15}N \rightarrow$ Si; δS only	882			

OXYGEN					O 8
	Stopping Power			Range	
Energy	Targets	Ref.	Energy	Targets	Ref.
20-140k	C	166	30-110k	He, O_2, H_2O	92
2-24M	C, Ar, Ni, Ag, Au	195	16-160M	Al	141
20-130k	Al	203	32-160M	O, Ni, Emulsion	220
32-160M	O, Ni, Emulsion	220	16-160M	Emulsion	214
25-50k	N_2	229	50-110M	Al, Cu, Au	213
81-479k	C	247	3-120M	Emulsion	253
0.36-3.2M	C, Al, Ni, Ag, Au	249	0.3-2.0M	Be	
5-30M	Ni, Ag, Au	264	0.2-1.5M	C	350
15-140k	B	266	0.6-11M	H_2, He, CH_4, Benzene, Air, Ar	362
19-40M	Au, Ag	322	8-160M	Emulsion	551
100-500k	Si (cryst.)	339	50k	Si	743
100-500k	Air	341	40k	Au	881
20-200k	N_2				
25-160k	Ar	342			
0.3-2M	Be				
0.2-1.5M	C	350			
0.56-11M	H_2, He, CH_4, Benzene, Air, Al, Ar, Ni, Ag, Au	362			
10-120k	CH_4	373			
150-500k	He; also δS	406			
200-500k	Ne, Air; δS only	409			
10-25M	Ar	421			
200-500k	Ne, He, Air; also δS	434			
8-66M	Ni, Ge, Y, Ag, Au	479			
22-46k	C	481			
0.2-2.0M	Cr, Fe, Co, Ni, Cu, Zn	485			
32-160M	Si	488			
50k	C				
2.9M	$O^- \rightarrow$ C, Ni				
5.8M	$O^-_2 \rightarrow$ C, Ni	532			
27.8-40M	Ag (cryst.); also δS, charge state dep.	573, 631			
7-38M	Au				
8-30M	Ni	575			
3-120k	CH_4	612			
7-35M	Ni, Au	627			
10M	An (cryst.); also δS	673			
2-20k	Ag	712			
16-160M	Mylar, polythene	722			
0.3-2M	Si; also δS, $\eta(\varepsilon)$	780			
0.3-1.7M	Al_2O_3; δS only	846			
26-46M	O				
52-92M	$O_2 \rightarrow$ C	869			
9-20M	Si; δS only	882			

FLUORINE			Range		F 9
Stopping Power			Range		
Energy	Targets	Ref.	Energy	Targets	Ref.
20-140k	C	166	19-190M	Al	141
		203	38-190M	O, Ni, Emulsion	220
38-190M	O, Ni, Emulsion	220	100-220M	Emulsion	300
138-473k	C	247	0.5-2.0M	Be	350
15-140k	B	266	3.9M	N_2	156
100-500k	Si (cryst.)	339	8-190M	Emulsion	551
30-140k	N_2		2-12M	Ni, Ta	688
25-60k	Ar	342	85k	Teflon	723
0.5-2.0M	Be	350	50k	Si	743
10-120k	CH_4	373	100-550k	Si	861
		612			
50k	C	488			
22-46k	C	479			
8-30M	Ni	575			
200-500k	Ne, He, Air; also δS	421			
3.8-66M	Ti, Fe, Ni, Cu, Ag, Au	821			

NEON			Range		Ne 10
Stopping Power			Range		
Energy	Targets	Ref.	Energy	Targets	Ref.
150-450k	H_2, He, Ar, Air	131			
20-140k	C	166			
20-130k	Al	203			
40-200M	O, Ni, Emulsion	220			
81-946k	C	247			
0.4-6.2M	C, Al, Ni, Ag, Au	250			
20-140k	B	266			
100-500k	Si (cryst.)	339			
100-500k	Air	341			
20-180k	N_2				
25-160k	Ar	342			
0.7-14M	H_2, He, CH_4, Benzene, Air, Ar, Al, Ni, Ag, Au	362			
10-120k	CH_4	373			
320-880k	Au (cryst.)	390			
0.2-10M	Ar	405			
200-400k	He; also δS				
	Ne, Air; δS only	406			
10-16M	Ar	409			
200-500k	He, Ne, Air; also δS	421			
22-46k	C	479			
50k	C	487			
		488			
18.5-19.8M	Al	489			
3-120k	CH_4	612			
1-10k	C	658			
100,200k	C	719			
20-200M	Mylar, Polythene	722			
0.3-3M	Si also; δS, $\eta(\varepsilon)$	780			
10,15M	C				
15M	He, N_2, SF_6, Ar; also δS	792			
2-60k	Si	871			
8-30M	$^{20}Ne \rightarrow$				
11-60M	$^{22}Ne \rightarrow$ Si; δS only	882			

SODIUM					Na 11
Stopping Power			**Range**		
Energy	Targets	Ref.	Energy	Targets	Ref.
20-70k	^{23}Na → C	166	1-100k	Al	143
25-70k	^{23}Na → Al	203	5-160k	^{24}Na → Al$_2$O$_3$	168
90-898k	^{23}Na → C	247	20-160k	^{24}Na → Al (cryst.)	174
25-70k	^{23}Na → B	266	24k	^{22}Na, ^{24}Na → Al	179
100-500k	Si (cryst.)	339	0.7-60k	^{24}Na → Al	180
100-500k	^{23}Na → Air	341	30k	^{24}Na → Al	181
3-20k	Cu	346	0.5-160k	^{24}Na → Al$_2$O$_3$, WO$_3$	182
0.85-16M	H$_2$, He, CH$_4$, Benzene, Air, Ar,		0.3-160k	^{24}Na → W (cryst.)	183
	Al, Ni, Ag, Au	362	0.1-1.5M	^{24}Na → W (cryst.)	
10-120k	^{23}Na → CH$_4$	373	0.1-0.5M	^{24}Na → Al (cryst.)	272
330-540k	^{23}Na → Au (cryst.)	390			286
50k	H$_2$, D$_2$, He, Ne, Ar; δS also	396	40k	Au, W (both cryst.)	
22-46k	C	479			304
50k	C	488	40-65k	Al (cryst.). Temp. dep.	
50-150k	C; dep. on scatt. angle	662	50k	^{24}Na → H$_2$, D$_2$, Ne, Ar	311
200-500k	He, Ne, Air; also δS	421	100-550k	^{24}Na → Au	338
			1335k	^{24}Na → Al	332
					364
			6-14k	Cu, Ag	359
			5-15k	Au	361
			0.85-16M	H$_2$, He, CH$_4$, Benzene, Air, Ar	362
			0.3-20M	Air	451
			40k	GaAs (cryst.)	22
			10-100k	^{24}Na → Si (cryst.)	27
			3.81M	Al	583
			2.4-6k	ZnS	584
			23k	Al, Si, W, Au (all cryst.)	613
			200k	Si (cryst.; chann., rand.)	616
			20,60k	SiO$_2$	621
			5-25k	Ni	676
			2-20k	SiC	682
			0.5-160k	Ta$_2$O$_5$	228
			40-1000k	Al$_2$O$_3$	357
			20-100k	^{23}Na → Ta	758
			0.43-1.58k	RbCl	794
			10-80k	^{24}Na → Al$_2$O$_3$, Ta$_2$O$_5$, WO$_3$	819
			40k	^{24}Na → W	881
			60k	Si	886

MAGNESIUM					Mg 12
Stopping Power			**Range**		
Energy	Targets	Ref.	Energy	Targets	Ref.
20-130k	^{24}Mg → C	166	328k	^{27}Mg → Al	332
135-766k	^{25}Mg → C	247	0.9-17M	H$_2$, He, CH$_4$, Benzene, Air, Ar	362
100-500k	^{24}Mg → Si (cryst.)	339	328k	^{27}Mg → Al	363
200-500k	^{24}Mg → Air	341	1.95M	Al	583
0.85-17M	^{24}Mg → H$_2$, He, CH$_4$, Benzene,				
	Air, Ar, Al, Ni, Ag, Au	362			
10-120k	^{24}Mg → CH$_4$	373			
370-560k	Au (cryst.)	390			
200-500k	He; also δS				
	Ne, Air; δS only	406			
	He, Ne, Air; also δS	421			
50k	C	488			
4.8-84M	Ti, Fe, Ni, Cu, Ag, Au	821			

ALUMINUM					Al 13
Stopping Power			Range		
Energy	Targets	Ref.	Energy	Targets	Ref.
25-80k	C	203	0.6k	$^{28}Al \rightarrow Al_2O_3$	332
88-875k	C	247	1-19M	H_2, He, CH_4, Benzene, Air	362
100-500k	Si (cryst.)	339	50k	C, Cu, Mo, Ta	461
200-400k	Air	341	60k	SiC	449
1-19M	H_2, He, CH_4, Benzene, Air, Ar,				450
	Al, Ni, Ag, Au	362			528
10-120k	CH_4	373			621
5.4-94M	Ti, Fe, Ni, Cu, Ag, Au	821	200k	Si	454
			2.7-40M	Mica	703
			60k	Si	807
			40k	Si; also $\delta R_\perp$	843
			50-120k	Ag	886

SILICON					Si 14
Stopping Power			Range		
Energy	Targets	Ref.	Energy	Targets	Ref.
0.2-3.1M	Si	186	10-28k	$^{30}Si \rightarrow Cu$	
20-130k	C	203	25k	$^{30}Si \rightarrow Ta$	200
21.2k-3.2M	Si	216			
430-560k	Si	245			
133-780k	C	247			
100-400k	Si	323			
100-500k	Si (cryst.)	339			
10-120k	CH_4	373			
0.3-2.0M	Si; also $\delta S, \eta(\varepsilon)$	710			
0.3-3.0M	Si; also $\delta S, \eta(\varepsilon)$	780			
68-157k	Si; also $\eta(\varepsilon)$	829			

PHOSPHORUS					P 15
Stopping Power			Range		
Energy	Targets	Ref.	Energy	Targets	Ref.
25-130k	C	203	40k	$^{32}P \rightarrow Si$ (cryst.)	243
137-849k	$^{31}P \rightarrow C$	247	0.1-1.5M	$^{32}P \rightarrow W$ (cryst.)	272
100-500k	$^{31}P \rightarrow Si$ (cryst.)	339			286
200-500k	$^{31}P \rightarrow Air$	341	10-70k	Si (cryst.)	333
1.1-22M	H_2, He, CH_4, Benzene, Air, Ar,		10-110k	$^{32}P \rightarrow Si$ (cryst.)	334
	Al, Ni, Ag, Au	362	150-500k	Au	338
10-120k	CH_4	373	10-150k	SiO_2	349
50k	C	488	1.1-22M	H_2, He, CH_4, Benzene, Air, Ar	362
32-46k	C	479	0.5-1.7M	Si (cryst.)	380

(CONT'D)

PHOSPHORUS (CONT'D)

				Energy	Targets	Ref.
				60k	Cu	415
				40k	NaCl (cryst.)	460
				100-300k	Si (cryst.)	455
				30-600k	$^{31}P \to$ Si (cryst.)	452
				40-120k	Si (cryst.)	27
				30-900k	Si (cryst.)	498
				40-250k	Si (cryst. and amorph.)	546
				70-280k	Si	552
				40k	^{31}P, $^{32}P \to$ Si (cryst.)	603
				50-145k	Si	638
				280k	$^{31}P \to$ Si	644
				1M	Si	650
				150k	Si (cryst.; chann., rand.)	671
				40k	InSb (cryst.; chann., rand.)	696
				16-60k	Si	702
				200k	(cryst.)	705
				40-120k	Si (cryst., amorph.)	706
				100-280k	$^{31}P \to$ Si	454
				15k	Si	831
				145,260k	$^{31}P \to$ Si; also $\delta R_\perp$	835
				100k	Si	858
				30k	Si	875
				40k	Au, W	881

SULPHUR   S 16

Stopping Power			Range		
Energy	Targets	Ref.	Energy	Targets	Ref.
25-125k	C	203	10-40k	GaAs	
168-753k	$^{32}S \to$ C	247	40k	GaP, Ge (all cryst.)	22
5-30M	Ni, Ag, Au	264	10-40k	Si (cryst.)	27
19-40M	Au, Ag	322	100k	$S_2 \to$ GaAs	530
200-500k	$^{32}S \to$ Air	341	20-40k	GaAs (cryst.; chann., rand.)	681
15-95M	Mylar	347			
5-90M	H_2, He, N_2, Ar, Kr	354			
10-120k	CH_4	373			
500-920k	Au (cryst.)	390			
9-30M	Au	575			
0.3-3M	Si; also δS, $\eta(\varepsilon)$	780			
6.4-112M	Ti, Fe, Ni, Cu, Ag, Au	821			
100M	Si (cryst.)	883			

CHLORINE — Cl 17

Stopping Power			Range		
Energy	Targets	Ref.	Energy	Targets	Ref.
4-40M	C, Ar, Ni, Ag, Au	195	1.2-24M	H_2, He, CH_4, Benzene, Air, Ar	362
20-125k	C	203	45k	$^{34}Cl \rightarrow$ Saran	723
134-1133k	$^{35}Cl \rightarrow$ C	247			
100-500k	$^{35}Cl \rightarrow$ Si (cryst.)	339			
200-500k	$^{35}Cl \rightarrow$ Air	341			
15-95M	$^{35}Cl \rightarrow$ Mylar	347			
5-90M	H_2, He, N_2, Ar, Kr	354			
1.2-24M	H_2, He, CH_4, Benzene, Air, Ar, Al, Ni, Ag, Au	362			
10-120k	CH_4	373			
670-970k	Au (cryst.)	390			
50k	H_2, D_2; also δS	396			
10-90M	Ni, Ge, Y, Ag, Au	434			
20-50k	H_2	652			
7-120M	Ti, Fe, Ni, Cu, Ag, Au	821			

ARGON — Ar 18

Stopping Power			Range		
Energy	Targets	Ref.	Energy	Targets	Ref.
80-400M	H_2, N_2, Ar	148	20-250k	H_2, He, Ar, Air	48
25-210k	C	203	50-160k	He, N_2, Ar, Air	57
25-50k	N_2	229	0.7-2250k	Al	157
0.03-2.9k	Ar	238	2-200k	W	162
138-1290k	C	247	50-500k	Be, B, C, Al	164
15-90k	ZnS:Ag	295	0.5-160k	$^{41}Ar \rightarrow Al_2O_3$, WO_3	182
10-100k	ZnO:Zn rel. to H^+ proj.	325	0.3-160k	$^{41}Ar \rightarrow$ W (cryst.)	183
100-500k	Si (cryst.)	339	1-30k	LiF, NaF, MgF_2, CaF_2, ZnS	191
150-1000k	Air	341	2.7k	W	199
10-120k	CH_4	373	40-400M	Emulsion	214
540-980k	Au (cryst.)	390	80-400M	H_2, N_2, Ar	208
10-27M	Ar	409	0.5-2M	Be, C	310
32-46k	C	479	5-60k	LiF, ZnS, CsJ	416
50k	C	488	4.2-5.6k	SiO_2, TiO_2	417
5-30k	SiO_2	610	284M	Polymetylmetacrylat	448
255M	Si; δS only	628	0.3-1.0M	N_2, O_2, Ar, Air	451
50-250k	C; dep. on scatt. angle	636	10-80k	CaF_2, RbJ	643
50-300k	C; dep. on scatt. angle	662	4-60M	Mica	703
100,200k	C	719	46-82k	Ge	769
0.3-3.0M	Si; also δS, $\eta(\varepsilon)$	780	20-160k	$^{41}Ar \rightarrow Al_2O_3$	
5-15M	N_2, C		20-80k	$^{41}Ar \rightarrow WO_3$	819
15M	SF_6, Ar; also δS	792	40k	Si; also $\delta R_\perp$	843
12-15M	Si; δS only	882			

POTASSIM					K 19
Stopping Power			**Range**		
Energy	Targets	Ref.	Energy	Targets	Ref.
25-65k	C	203	30k	$^{42}K \rightarrow Al$	180
138-1138k	$^{39}K \rightarrow C$	247	6-14k	Cu, Ag	240
100-500k	$^{39}K \rightarrow Si$ (cryst.)	339	0.1-1.5M	$^{42}K \rightarrow W$ (cryst.)	
200-800k	$^{39}K \rightarrow Air$	341	0.1-0.5M	$^{42}K \rightarrow Al$ (cryst.)	272
3-20k	Cu	346			286
1.4-27M	H_2, He, CH_4, Benzene, Air, Ar,		0.07-1.0M	$^{42}K \rightarrow W$ (cryst.)	302
	Al, Ni, Ag, Au	362	150-500k	$^{42}K \rightarrow Au$	338
10-120k	CH_4	373	40k	$^{42}K \rightarrow W$ (cryst.)	336
340-540k	Au (cryst.)	390	1.4-27M	H_2, He, CH_4, Benzene, Air, Ar	362
50k	D_2, He, Ne, also δS	396	1.0-1.7M	$^{43}K \rightarrow Ar$	156
1-20k	C	659	50k	Si; also δR, δR	540
			2.4-6k	ZnS	584
			40k	Si, Al, W, Au (all cryst.)	613
			55k	$^{42}K \rightarrow$ Cu, Ag, Au, V, Mo, Nb,	
				Ta, W (all cryst.)	630
			3-20k	SiC	682
			0.6-160k	$^{42}K \rightarrow Ta_2O_5$	228
			40-1000k	$^{42}K \rightarrow Al_2O_3$	357
			20-80k	$^{42}K \rightarrow Ta_2O_5$	819

CALCIUM					Ca 20
Stopping Power			**Range**		
Energy	Targets	Ref.	Energy	Targets	Ref.
191-874k	$^{40}Ca \rightarrow C$	247	4-60M	Mica	703
250-500k	$^{40}Ca \rightarrow Air$	341	121-335k	$^{40}Ca \rightarrow C$	860
10-120k	CH_4	373			
2-9M	Ca	393			
50k	H_2	652			

SCANDIUM					Sc 21
Stopping Power			**Range**		
Energy	Targets	Ref.	Energy	Targets	Ref.
200-1200k	C	294	k	Al	260
		315	2.3-3.5M	$^{47}Sc \rightarrow {}^{51}V$	
10-120k	CH_4	373	3.0-3.7M	$^{46}Sc \rightarrow {}^{51}V$	309
200-500k	Air	341			
370-950k	Au (cryst.)	390			

TITANIUM					Ti 22
Stopping Power			Range		
Energy	Targets	Ref.	Energy	Targets	Ref.
200-1000k	C	294	k	Al	260
		315	33k	^{45}Ti → Ti	723
150-500k	Air	341			
10-120k	CH_4	373			

VANADIUM					V 23
Stopping Power			Range		
Energy	Targets	Ref.	Energy	Targets	Ref.
			3.6-4.9M	^{48}V → ^{51}V	309

CHROMIUM					Cr 24
Stopping Power			Range		
Energy	Targets	Ref.	Energy	Targets	Ref.
0.4-1.485M	^{52}Cr → C	294	k	Al	260
		315	0.1-1.5M	^{51}Cr → W (cryst.)	272
300-800k	^{52}Cr → Air	341	3.0-5.1M	^{51}Cr → ^{51}V	309
0.02-0.09k	V (cryst.)	596	5-80M	Mica	703
0.02-0.09k	$V_1V_2O_5$, VOC_2H_4,		5k	Ge	852
	$V(CH(COCH_3)_2)_3$				
	$V(CH(COCH_3)_2)_2$	830			

MANGANESE					Mn 25
Stopping Power			Range		
Energy	Targets	Ref.	Energy	Targets	Ref.
0.3-1.2M	^{55}Mu → C	294	k	Al	260
		315	1.1-2.0M	^{54}Mn → V	
380-660k	Au (cryst.)	390	2.2-3.8M	^{55}Mn → V	309
50k	Ar; also δS	396	219k	^{56}Mn → Fe	332
100,200k	C	719			367

IRON					Fe 26
Stopping Power			Range		
Energy	Targets	Ref.	Energy	Targets	Ref.
0.3-1.5M	$^{56}Fe \rightarrow C$	294 315	k	Al	260
430-970k	Au (cryst.)	390	30k	$^{53}Fe \rightarrow Fe$	723
40,50k	H_2	652			

COBALT					Co 27
Stopping Power			Range		
Energy	Targets	Ref.	Energy	Targets	Ref.
0.3-1.2M	$^{59}Co \rightarrow C$	294 315	0.5-4.5M	Al	254
			k	Al	260
			0.2-5M	Fe	299
			0.17-0.65M	$^{55}Co \rightarrow$	
			0.22-0.70M	$^{56}Co \rightarrow$	
			0.57-0.69M	$^{57}Co \rightarrow$	
			0.30-0.68M	$^{58}Co \rightarrow Ni$	
			2.2-5.2M	$^{56}Co \rightarrow$	
			1.4-2.9M	$^{57}Co \rightarrow ^{54}Fe$	309
			145k	$^{58}Co \rightarrow Ni$	335
			40k	$^{58}Co \rightarrow Au, W$	881
			60k	$^{57}Co \rightarrow Cu$	415

NICKEL					Ni 28
Stopping Power			Range		
Energy	Targets	Ref.	Energy	Targets	Ref.
12.5M	Ni	708	0.5-4.5M	Al	254
48M	C; δS only	915	k	Al	260
			0.2-5M	Fe	294
			0.57-0.67M	$^{57}Ni \rightarrow Ni$	
			1.4-2.9M	$^{57}Ni \rightarrow ^{54}Fe$	309
			6-85M	Mica	703

COPPER					Cu 29
Stopping Power			Range		
Energy	Targets	Ref.	Energy	Targets	Ref.
0.4-1.5M	$^{65}Cu \rightarrow C$	294 315	3.35-17.6M k 0.1-1.5M 0.5k 60k 10-40k 100k - 2.4-57.5k 25k 200k 40k 35k	^{60}Cu, $^{61}Cu \rightarrow Al$ Al $^{64}Cu \rightarrow W$ (cryst.) $^{64}Cu \rightarrow CuO$ $^{67}Cu \rightarrow Cu$ $^{64}Cu \rightarrow Si$ (cryst.) Be - Cu ^{62}Cu, ^{64}Cu Al, Cu Si; also $\delta R_\perp$ Si; $\delta R_\perp$ only	242 260 272 332 415 27 536 639 563 723 826 843 887

ZINC					Zn 30
Stopping Power			Range		
Energy	Targets	Ref.	Energy	Targets	Ref.
50k	H_2, D_2, He, Ne, Ar; also δS	396	140,280k 280k 260k 25k 40k	SiO_2 Si_3N_4 Al_2O_3 $^{63}Zn \rightarrow Zn$ $^{65}Zn \rightarrow W$	13 539 723 881

GALLIUM					Ga 31
Stopping Power			Range		
Energy	Targets	Ref.	Energy	Targets	Ref.
20-50k	H_2	652	0.6-1.2M 1.0-1.3M 70-1000k 50k 56k 260k 140,280k 40k 60k	H_2, D_2, He, N_2, Ar Cu Cu, Zn $^{66}Ga \rightarrow H_2$, D_2, ^3He, ^4He, N_2, Ar Si (cryst.) Si_3N_4, Al_2O_3 ^{69}Ga, $^{71}Ga \rightarrow Si$ Si (cryst.) $^{69}Ga \rightarrow Si$	156 178 311 348 13 454 713 842

GERMANIUM					Ge 32
Stopping Power			Range		
Energy	Targets	Ref.	Energy	Targets	Ref.
20-100k	Ge	226	k	Al	260
		283	20-100k	Ge	283
21-997k	Ge; $\eta(\varepsilon)$	246	10-35k	Si	837
0.4-1.5M	^{74}Ge → C	294	40k	Si	845
		315			
50k	H_2	652			
25k	Ge; $\eta(\varepsilon)$	664			
1-1.8k	Ge; $\eta(\varepsilon)$	672			
19-36k	Ge; $\eta(\varepsilon)$	829			
10-30k	Ge; $\eta(\varepsilon)$	840			

ARSENIC					As 33
Stopping Power			Range		
Energy	Targets	Ref.	Energy	Targets	Ref.
			10-70k	Si (cryst.)	333
			1.0-1.7M	Si (cryst.)	380
			120k	Si	12
			300k	SiO_2	
			260k	Al_2O_3	13
					539
			40-150k	SiO_2	518
			45k	Si	527
			35-130k	Si	544
			40k	Si	545
			80-480k	Si	552
			150-200k	Si (cryst.)	615
			50-250k	Si	640
			280k	Si	454
			45k	Si	716
			25,40k	Si	744
			25k	Si	745
			10-35k	Si	837
			5-60k	Si	845

SELENIUM					Se 34
Stopping Power			Range		
Energy	Targets	Ref.	Energy	Targets	Ref.
			150-280k	SiO_2	
			200k	Si_3N_4	
			200k	Al_2O_3	13
					539
			8-120M	Mica	703
			200k	Al	826
			40k	^{75}Se → W	881

BROMINE					Br 35
Stopping Power			**Range**		
Energy	Targets	Ref.	Energy	Targets	Ref.
10-100M	Be, C, Al, Ni, Ag, Au	270	30-260M	^{80}Br → Emulsion	231
20-100M	UF_4	285	0.1-1.5M	^{82}Br → W (cryst.)	272
0.6-1.5M	^{79}Br → C	294	100-165M	Emulsion	300
		315	30k	^{82}Br → Ta_2O_5, ZrO_2	325
30-90M	^{79}Br → Mylar	347	2.8-56M	H_2, He, CH_4, Benzene, Air, Ar	362
5-90M	^{79}Br → H_2, He, N_2, Ar, Kr	354			
2.8-56M	H_2, He, CH_4, Benzene, Air, Al, Ar, Ni, Ag, Au	362			
20-80M	^{79}Br → Au (cryst.); also δS	593			
12-70M	^{79}Br → Au (cryst.); also δS	777			

KRYPTON					Kr 36
Stopping Power			**Range**		
Energy	Targets	Ref.	Energy	Targets	Ref.
0.6-1.5M	^{86}Kr → C	294	40k	WO_3	162
		315	2-600k	Al, W	158
25-90k	^{86}Kr → ZnS:Ag	295	10-150k	Cu	161
10-100k	^{84}Kr → ZnO:Zn rel. to H^+	325	40k	Al	163
2.9-59M	H_2, He, CH_4, Benzene, Air, Al, Ar, Ni, Ag, Au	362	50-500k	Be, B, C, Al	164
400-770k	Au (cryst.)	390	5-160k	Al_2O_3	168
100,200k	C	719	20-160k	Al (cryst.)	174
90M	Si		0.5-160k	Al_2O_3, WO_3	182
83M	Ar; both δS only	863	0.3-160k	W (cryst.)	183
180-300M	Nd, Dy, Er	864	25-125k	Cu	184
80-100M	C; also δS	867	80k	GaAs (cryst.)	188
0.2-5M	Ag (cryst.)	870	2.7k	W	199
			70k	W (cryst.)	237
					265
			70k	Au (cryst.)	271
			0.1-1.5M	W (cryst.)	272
					286
			5k	W (cryst.)	297
			40k	Au (cryst.). Temp. dep.	304
			0.5-2.0M	Be, C, Al	310
			30k	Ta_2O_5, ZrO_2	328
			2.9-59M	H_2, He, CH_4, Benzene, Air, Ar	362
			3-9k	Pt, Al, KCl	377
			10-40k	GaAs (cryst.)	378
			100-400k	Ag	
			30-60k	Au	415
			5-60k	LiF, ZnS, CsJ	416
			30-60k	NaCl (cryst.)	460
			40k	GaAs (cryst.)	22
			10-40k	Si (cryst.)	27
			180k	Si (amorph.); also δR, $\delta R_{trans.}$	125
					510
			100,180k	Si (amorph.); also δR, $\delta R_{trans.}$	540
			20-80k	Al, Si	611

(CONT'D)

KRYPTON (CONT'D)

		20-150k	Al (cryst.)		
		80k	Si, W, Au (all cryst.)		613
		10-80k	CaF_2, RbJ		643
		10k	Mo		678
		8-120M	Mica		703
		40-1000k	Al_2O_3		357
		10k	KCl, TiO_2, Al_2O_3		759
		20-80k	$^{85}Kr \rightarrow Al_2O_3$, Ta_2O_5, WO_3		819
		40k	Si; also $\delta R_\perp$		843
		40k	$^{85}Kr \rightarrow W$		881
		27k	W (cryst.)		884
		45k	Si; $\delta R_\perp$ only		887
		30k	Be		900

RUBIDIUM — Rb 37

Stopping Power			Range		
Energy	Targets	Ref.	Energy	Targets	Ref.
9-20k	Cu		20-160k	$^{86}Rb \rightarrow$ Al (cryst.)	174
12-17k	Ag, Au	641	30k	$^{86}Rb \rightarrow$ Al	181
90M	Si; δS only	863	0.1-1.5M	$^{86}Rb \rightarrow$ W (cryst.)	272
80-100M	C; also δS	867	30k	$^{86}Rb \rightarrow Ta_2O_5$, ZrO_2	328
			25-125k	Si	649
			0.5-160k	Ta_2O_5	228
			1.58-3.04k	RbCl	794
			250k	Al, Cu	826

STRONTIUM — Sr 38

Stopping Power			Range		
Energy	Targets	Ref.	Energy	Targets	Ref.
k	Al	260			
45k	Si (cryst.; chann, random)	796			
80-100M	C; δS also	867			

YTTRIUM					Y 39
Stopping Power			Range		
Energy	Targets	Ref.	Energy	Targets	Ref.
0.5-1.4M	C	294	k	Al	260
		315	2.6-3.0M	^{86}Y →	
630-940k	Au (cryst.)	390	1.8-3.2M	^{88}Y → ^{90}Zr, Nat. Zr	309
50k	H$_2$, D$_2$, He, Ne; also δS	396			
84M	C; δS only	863			
80-100M	C; also δS	867			

ZIRCONIUM					Zr 40
Stopping Power			Range		
Energy	Targets	Ref.	Energy	Targets	Ref.
30-50k	H$_2$	652	k	Al	260
84M	C; δS only	863	97M	^{95}Zr → Mica, Al (cryst.)	262
80-100M	C; also δS	867	2.7-3.2M	^{89}Zr →	
			1.7-3.5M	^{88}Zr →	
			2.5-3.0M	^{87}Zr →	
			2.8-3.0M	^{86}Zr → ^{90}Zr, Nat. Zr	309
			97M	^{95}Zr → Al	827

NIOBIUM					Nb 41
Stopping Power			Range		
Energy	Targets	Ref.	Energy	Targets	Ref.
80-100M	C; also δS	867	2.2-3.1M	^{87}Nb → ^{90}Zr; Natl. Zr	309
			143k	Nb	609
					791
			97M	^{95}Nb → Al	827

MOLYBDENUM					Mo 42
Stopping Power			Range		
Energy	Targets	Ref.	Energy	Targets	Ref.
280-520k	Au (cryst.)	390	k	Al	260
			2.2-3.1M	^{90}Mo → ^{90}Zr, Nat. Zr	309
			179.4M	^{99}Mo → Al, Ni, Cu, Pb	699
			16k	^{91}Mo → Mo	723
			150k	Al, Cu	826

RUTHENIUM					Ru 44
Stopping Power			Range		
Energy	Targets	Ref.	Energy	Targets	Ref.
			300k	Cu	826

RHENIUM					Rh 45
Stopping Power			Range		
Energy	Targets	Ref.	Energy	Targets	Ref.
				Al	260

PALLADIUM					Pd 46
Stopping Power			Range		
Energy	Targets	Ref.	Energy	Targets	Ref.
			49,91M	Al Nd	260 895

SILVER					Ag 47
Stopping Power			Range		
Energy	Targets	Ref.	Energy	Targets	Ref.
390-760k 50k 100-300k	Au (cryst.) H_2, D_2, He, Ne, Ar; also δS ^{107}Ag → H_2; also δS	390 396 406	32-320M k 2.9-27.2k 11-160M 20-140k 14k 200k 40k	^{108}Ag → Emulsion Al Ag Mica ^{107}Ag → Si, SiO_2 ^{106}Ag → Ag Al ^{110m}Ag → W	231 260 563 703 718 723 826 881

CADMIUM					Cd 48
Stopping Power			Range		
Energy	Targets	Ref.	Energy	Targets	Ref.
490-970k 18M	Au (cryst.) Cd	390 708	k 260k 300k 40k 40k	Al SiO_2, Si_3N_4, Al_2O_3 Al Si; also $\delta R_\perp$ Si; $\delta R_\perp$ only	260 13 539 826 843 887

INDIUM					In 49
Stopping Power			Range		
Energy	Targets	Ref.	Energy	Targets	Ref.
			56k 20-80k	Si (cryst.) Si	348 753

TIN					Sn 50
Stopping Power			Range		
Energy	Targets	Ref.	Energy	Targets	Ref.
			k	Al	260

ANTIMONY					Sb 51
Stopping Power			Range		
Energy	Targets	Ref.	Energy	Targets	Ref.
390-780k 30-50k 80-100M	Au (cryst.) H_2 C; also δS	390 652 867	0.1-1.5M 20k 120,260k 100k 20k 40k 5-60k	$^{122}Sb \rightarrow$ W (cryst.) Si (cryst.) ^{121}Sb, $^{123}Sb \rightarrow$ Si Si Si Diamond Si	272 301 454 659 753 810 845

TELLURIUM					Te 52
Stopping Power			Range		
Energy	Targets	Ref.	Energy	Targets	Ref.
80-100M	C; also δS	867	280k 260k	SiO_2 Si_3N_4, Al_2O_3	13 539

IODINE					I 53
Stopping Power			Range		
Energy	Targets	Ref.	Energy	Targets	Ref.
25-115M	C, Al, Ni, Au	170	300-235M	Emulsion	300
40M	Au (cryst.)	258	40k	^{131}I → W	881
60M	Au (cryst.); also δS	267			
		384			
10-100M	Be, C, Al, Ni, Ag, Au	270			
		293			
20-100M	UF_4	285			
90-200M	Be, C, Al, Ni, Ag, Au, UF_4	289			
		293			
60-105M	Mylar	347			
5-90M	H_2, He, N_2, Ar, Kr	354			
21.6-60M	Au, Ag (cryst.); also δS	483			
15-60M	Au (cryst.); also δS	21			
21.6-31.3M	Ag (cryst.); also δS	572			
		576			
15,21.6M	Au (cryst.); also δS	673			
23-82M	Au (cryst.); also δS	777			
0.6-60M	Ag (cryst.)	870			
22-115M	C, Al, Ni, Au	909			
110M	C; δS only	915			

XENON					Xe 54
Stopping Power			Range		
Energy	Targets	Ref.	Energy	Targets	Ref.
350-760k	Au (cryst.)	390	0.7-240k	Al	157
300-700k	C	624	2-40k	^{125}Xe → Al, W, Au	160
100,200k	C	719	2-200k	W	162
90-300k	C	720	50-500k	Be, B, C, Al	164
			5-160k	Al_2O_3	168
			40-80k	Si	169
			1-40k	^{125}Xe → W	172
			20-160k	^{125}Xe → Al (cryst.)	174
			5-80k	^{125}Xe → Si (cryst.)	173
			0.5-160k	Al_2O_3, WO_3	182
			0.3-160k	^{125}Xe, ^{133}Xe → W (cryst.)	183

(CONT'D)

XENON (CONT'D)

			Energy	Targets	Ref.
			40k	^{133}Xe → Au (cryst.) 4.2K	187
			5-30k	Ni, Mo, Cu	201
			40, 125k	^{133}Xe → W, WO$_3$	202
			40k	^{133}Xe → Ta$_2$O$_5$, ZrO$_2$, UO$_2$	241
			20k	^{133}Xe → W (cryst.)	263
			0.1-1.0M	^{133}Xe → W (cryst.)	
			0.04-0.5M	^{133}Xe → Al (cryst.)	272
			40k	NaCl, KBr, MgO, SiO$_2$, UO$_2$ (all cryst.)	276
			0.1-1.5M	^{133}Xe → W (cryst.)	286
			20-80k	Au (cryst.) Temp. dep.	303
			40k	Al, W (cryst.) Temp. dep.	304
			0.5-2.0M	Be, C, Al, V, Ni, Cu	310
			40k	^{133}Xe → Ta, W, Al, Cu, Au, Ir (all cryst.)	335
			30k	^{125}Xe, ^{133}Xe → Ta$_2$O$_5$, ZrO$_2$	328
			20-100k	Cu (cryst.)	494
			40-80k	Au	495
			20-80k	Al, Si	611
			20-150k	Au, W (cryst.)	
			20-80k	Si (cryst.)	
			80k	Al (cryst.)	
			0.5-100k	Ta$_2$O$_5$	
			40k	UO$_2$	613
			20-80k	Cu (cryst.)	660
			0.5-60k	Ta$_2$O$_5$	228
			40-1000k	Al$_2$O$_3$	357
			75k	^{129}Xe → Al, Al$_2$O$_3$	715
			20-150k	Si	751
			20-100k	Cu (cryst.)	766
			20-160k	^{125}Xe → Al$_2$O$_3$, Ta$_2$O$_5$, WO$_3$	819
			40k	^{133}Xe → W, ^{125}Xe → WO$_3$	881
			40k	Si; $\delta R_\perp$ only	887

CESIUM					Cs 55
Stopping Power			Range		
Energy	Targets	Ref.	Energy	Targets	Ref.
11-19k	Cu		2-50k	^{137}Cs → Al	181
16-20k	Ag, Au	641	4k	Ge	197
					198
			40,125k	^{134}Cs → W (cryst.), WO$_3$	202
			40k	^{134}Cs → ZrO$_2$	241
			4-20k	^{137}Cs → Ge	559
			2.4-6k	ZnS	584
			20-80k	Al, Si	611
			125k	Si, Al, W, Au (all cryst.)	613
			10-80k	CaF$_2$, RbJ	643
			5-50k	Si	648
			4-9k	W	701
			100k	Al	707
			0.5-160k	Ta$_2$O$_5$	228
			75k	^{133}Cs → Al	715
			10-100k	Si	795
			250k	Cu	
			300k	Al	826
			40k	^{134}Cs → W	881
			20,30k	^{133}Cs → Si	
			30k	^{133}Cs → Al	888
			20k	^{133}Cs → Cu, Al	891

BARIUM					Ba 56
Stopping Power			Range		
Energy	Targets	Ref.	Energy	Targets	Ref.
50k	H_2, D_2, He, Ne; also δS	396	2.8-14.2M	^{126}Ba, $^{128}Ba \rightarrow Al$	256
			65M	$^{140}Ba \rightarrow$ Mica, Al (cryst.)	262
			100k	Al	606
					707
			188M	$^{140}Ba \rightarrow$ Al, Ni, Cu	699
			100M	$^{140}Ba \rightarrow W$	760
			45k	Si (cryst.; chann., random)	796
			68.5,140M	Al (cryst.; chann., random; poly-cryst.)	804
			140M	$^{140}Ba \rightarrow Al$	827

LANTHANUM					La 57
Stopping Power			Range		
Energy	Targets	Ref.	Energy	Targets	Ref.
			100k	Al	606
			100k	Al	707
			140M	$^{140}La \rightarrow Al$	827

CERIUM					Ce 58
Stopping Power			Range		
Energy	Targets	Ref.	Energy	Targets	Ref.
			100k	Al	606
			45k	Si (cryst. chann., random)	796

PRASEODYMIUM					Pr 59
Stopping Power			Range		
Energy	Targets	Ref.	Energy	Targets	Ref.
			100k	Al	606
			37M	Ni, Ta	895

NEODYMIUM					Nd 60
	Stopping Power			Range	
Energy	Targets	Ref.	Energy	Targets	Ref.
70-250M	Nd	864	100k 45k	Al Si (chann., random)	606 796

SAMARIUM					Sm 62
	Stopping Power			Range	
Energy	Targets	Ref.	Energy	Targets	Ref.
			2-12M 100k 100k	^{142}Sm → Al Al ^{152}Sm → Si	255 606 707 873

EUROPIUM					Eu 63
	Stopping Power			Range	
Energy	Targets	Ref.	Energy	Targets	Ref.
			100k 80k 45k 300k 100k	Al ^{153}Eu → Al, Al$_2$O$_3$ ^{145}Eu → Si (cryst. chann., ran- dom) Cu ^{153}Eu → Si	707 715 796 826 873

GADOLINIUM					Gd 64
	Stopping Power			Range	
Energy	Targets	Ref.	Energy	Targets	Ref.
50k 100-500k	H$_2$, D$_2$, Ne; also δS H$_2$; also δS	396 663	k 50-150k 45k 50-150k 150k 100k	Al Nb Si (cryst. chann., random) Nb Al ^{157}Gd → Si	260 531 796 803 826 873

TERBIUM					Tb 65
Stopping Power			Range		
Energy	Targets	Ref.	Energy	Targets	Ref.
			4-29M	^{149}Tb → Al	557
			6-21M	H_2, D_2, Al	204
			100k	Al	707
			45k	Si (cryst. chann., random)	796
			100k	^{159}Tb → Si	873

DYSPROSIUM					Dy 66
Stopping Power			Range		
Energy	Targets	Ref.	Energy	Targets	Ref.
70-250M	Dy	864	6-21M	He, N_2, Ne, Ar, Kr, Xe	185
			6-21M	H_2, D_2, Al	204
			20-80k	Al, Si	611
			40k	Si; also $\delta R_\perp$	843
			100k	^{164}Dy → Si	873
			40k	Si; $\delta R_\perp$ only	887

ERBIUM					Er 68
Stopping Power			Range		
Energy	Targets	Ref.	Energy	Targets	Ref.
70-250M	Er	864	45k	Si (cryst. chann., random)	796

THULIUM					Tm 69
Stopping Power			Range		
Energy	Targets	Ref.	Energy	Targets	Ref.
			45k	Si (cryst. chann., random)	796

YTTERBIUM — Yb 70

Stopping Power			Range		
Energy	Targets	Ref.	Energy	Targets	Ref.
			k	Al	260

LUTETIUM — Lu 71

Stopping Power			Range		
Energy	Targets	Ref.	Energy	Targets	Ref.
50k	D_2, He; also δS	396	45k	Si (cryst., chann., random)	796

HAFNIUM — Hf 72

Stopping Power			Range		
Energy	Targets	Ref.	Energy	Targets	Ref.
50k	H_2, D_2, He, Ne; also δS	396	40k	^{181}Hf → W	881

TANTALUM — Ta 73

Stopping Power			Range		
Energy	Targets	Ref.	Energy	Targets	Ref.
10-140M	C, Al, Ag, Au	394	k	Al	260
			100-180M	Emulsion	300
			200k	Cu	826

TUNGSTEN					W 74
Stopping Power			Range		
Energy	Targets	Ref.	Energy	Targets	Ref.
			1.6-127k	^{187}W → W	165
			800k	Cu	327
			40k	Si; also $\delta R_\perp$	843

GOLD					Au 79
Stopping Power			Range		
Energy	Targets	Ref.	Energy	Targets	Ref.
100M	C; δS only	915	k	Al	260
			20-80k	Au (cryst.)	287
					613
			40-94k	Au (cryst.) Temp. dep.	304
			50k	^{198}Au → H$_2$, D$_2$, Ne, Ar	311
			0.05k	D$_2$, He, Ne, Ar, Xe	458
			100k	Be	529
			6.1-15.1k	Au	563
			71k	Au	609
			20-80k	Si, Al	611
			5,40k	Ta$_2$O$_5$	686
			100k	Al	707
			80k	^{197}Au → Al, Al$_2$O$_3$	715
			9k	^{196}Au → Au	723
			71k	^{196}Au → Au	791
			150k	Cu	
			200k	Al	826
			60k	Si, Al	845
			50k	Al	914

MERCURY					Hg 80
Stopping Power			Range		
Energy	Targets	Ref.	Energy	Targets	Ref.
50k	^{198}Hg → H$_2$, D$_2$, He, Ne, Ar; also δS	396	185k	Si, Al, W, Au (all cryst.)	613
100-500k	^{202}Hg → H$_2$, He; also δS	406			
100-500k	^{202}Hg → H$_2$; also δS	663			

THALLIUM — Tl 81

Energy	Stopping Power Targets	Ref.	Energy	Range Targets	Ref.
116k	^{208}Tl → Au (cryst.)	351	0.37-0.62M	^{200}Tl →	
100-500k	H_2; also δS	663	0.41-0.82M	^{199}Tl →	
			0.47-1.2M	^{198}Tl →	
			0.88-1.46M	^{197}Tl →	
			1.03-1.49M	^{196}Tl → ^{197}Au	309
			5,40k	Ta_2O_5	686
			0.5-160k	Ta_2O_5	228
			100k	^{205}Tl → Al, Al_2O_3	715
			40k	^{204}Tl → Au	881

LEAD — Pb 82

Energy	Stopping Power Targets	Ref.	Energy	Range Targets	Ref.
169k	^{208}Pb → C, Ag, Au	261	103k	^{206}Pb → Ar	230
169k	^{208}Pb → C	344	103k	^{206}Pb → Ar, He	239
169k	^{208}Pb →		103k	^{206}Pb → Ne, Ar, Xe, N_2, Air,	
146k	^{210}Pb → Au (cryst.)	351		Hydrocarbon gases	331
169k	^{208}Pb → Au	397	20-80k	Si, Al	611
25-75k	He	524	40k	Si	684
30-80k	H_2	652	10-80k	Si; also $\delta R_\perp$	843
100-500k	H_2; also δS	663	20k	Si	852
169k	^{208}Pb → Formvar	724	10-40k	Si; $\delta R_\perp$ only	887
20-156	H_2	876			

BISMUTH — Bi 83

Energy	Stopping Power Targets	Ref.	Energy	Range Targets	Ref.
50k	^{209}Bi → H_2, D_2, He, Ne, Ar; also δS	396	25k	CdS (cryst.)	496
					697
			20-80k	Si, Al	611
			240k	Si	454
			150k	Al, Cu	826
			2-60k	^{209}Bi → Si	842
			50-400k	Si; also $\delta R_\perp$	843
			1-40k	Si	
			1k	Ge	845
			45-400k	Si; $\delta R_\perp$ only	887
			200k	Fe, Ni (cryst., chann. and random)	919

POLONIUM					Po 84
Stopping Power			Range		
Energy	Targets	Ref.	Energy	Targets	Ref.
			140-280k	Bi	144
			6-12M	Al, Al_2O_3	330
			4-15M	Al	
			4-9M	Au	557
			25k	$^{210}Po \rightarrow$ CdS (cryst.)	749

ASTATINE					At 85
Stopping Power			Range		
Energy	Targets	Ref.	Energy	Targets	Ref.
			6-12M	Al, Al_2O_3	330
			4-15M	Al	
			4-9M	Au	557
			3.2-4.2M	Al	685

RADON					Rn 86
Stopping Power			Range		
Energy	Targets	Ref.	Energy	Targets	Ref.
			2-450k	Al, W	154
			0.1-1.5M	$^{222}Rn \rightarrow$ W (cryst.)	272
			140-210k	Al	
			70-210k	W	
			133k	Ag, Au	298
			0.5-160k	Ta_2O_5	228

FRANCIUM					Fr 87
Stopping Power			Range		
Energy	Targets	Ref.	Energy	Targets	Ref.

RADIUM					Ra 88
	Stopping Power			Range	
Energy	Targets	Ref.	Energy	Targets	Ref.
			96.8k	^{224}Ra → H$_2$, D$_2$, He, N$_2$, Ne, Ar	144
			96.8k	^{224}Ra → Ar, Ne, O$_2$, N$_2$, CH$_4$, He, H$_2$	296
					332
			96.8k	^{224}Ra → H$_2$, He, N$_2$, O$_2$, Ne, Ar, Kr, Xe	446
			96.8k	^{224}Ra → H$_2$, D$_2$, T$_2$, ^{3}He, ^{4}He, ^{14}N$_2$, ^{15}N$_2$, ^{16}O$_2$, ^{18}O$_2$, ^{20}Ne, ^{22}Ne	447

THORIUM					Th 90
	Stopping Power			Range	
Energy	Targets	Ref.	Energy	Targets	Ref.
			725k	H$_2$, D$_2$, He, N$_2$, Ar	145
			725k	D$_2$, He, N$_2$, Ar	144
			k	Al	260

URANIUM					U 92
	Stopping Power			Range	
Energy	Targets	Ref.	Energy	Targets	Ref.
30-90M	C, Al, Ni, Ag, Au	477	100-240M	Emulsion	300
5-110M	He, Air, Kr	552			
50k	H$_2$	652			

NEPTUMIUM					Np 93
	Stopping Power			Range	
Energy	Targets	Ref.	Energy	Targets	Ref.
92k	^{237}Np → Collodium	400	924k	^{239}Np → Ar	230

CURIUM					Cm 96
	Stopping Power			Range	
Energy	Targets	Ref.	Energy	Targets	Ref.
			60-400k	$^{240}Cm \rightarrow ^{241}AmO_2$	525

MESONS, HYPERONS, ETC.					
	Stopping Power			Range	
Energy	Targets	Ref.	Energy	Targets	Ref.
370M	$\pi^- \rightarrow$ Si	314	G	$\mu \rightarrow$ Pb	196
0.5-10.5G	$\mu^+, \mu^- \rightarrow$ NaI; also δS	321	4.12M	$\mu^+, \mu^- \rightarrow$ He	418
0.3-120G	μ^+ Scint. Plast.	352	1-10M	$\Sigma^- \rightarrow$ Emulsion	444
		360	0.85-1.1G	$\mu^+, \mu^- \rightarrow$ Fe	462
4-25M	$\pi^+, \pi^- \rightarrow$ Emulsion	383	100,500M	$\pi \rightarrow$	
50-200M	$\pi^+ \rightarrow$ Si; δS only	385	810M	$\mu \rightarrow$ Emulsion	43
0.2-1.6G	$\pi^- \rightarrow$ Emulsion	34	36.6M	$\mu^+ \rightarrow$	
515M	$\pi^- \rightarrow$ Emulsion rel. to electrons	38	200-700M	$\pi^+ \rightarrow$ Emulsion	585
0.5-10G	$\mu^+ \rightarrow$ Plast. scint.	41	12M	$\Sigma^+, \Sigma^- \rightarrow$ Emulsion	345
$0.79 < \beta <$	$\pi, \mu \rightarrow$ Emulsion		9-25M	$\mu^+ \rightarrow$ Al, Steel, Polyethylene	635
0.94		49	39.3M	$\mu^+ \rightarrow$ Al	
0.39-2.2G	$\mu^+ \rightarrow$ Xylene rel. to min.	52	33.0,37.6M	$\pi^+, \pi^- \rightarrow$ Al	775
61-222M	$\pi \rightarrow$				
5.23M	$\mu \rightarrow$ NaI	55			
5G	$\pi \rightarrow$ Emulsion, rel. to min.	58			
0.4-100G	$\mu^+ \rightarrow$ Plast. Scint.	68			
1.5-100G	$\mu \rightarrow$ Ne	69			
24-224M	$\pi \rightarrow$ Emulsion rel. to min.	72			
0.5-30G	$\mu \rightarrow$ O_2	73			
		74			
0.2-20G	$\mu, \pi \rightarrow$ Air rel. to min.	79			
0.6-1.3G	$\mu^+ \rightarrow$ Ar; also δS	80			
0.3-30G	$\mu \rightarrow$ Ne + CH_4	86			
0.14-140G	$\pi, \mu \rightarrow$ Emulsion rel. to 5.2 GeV π	94			
0.3-12G	$\mu \rightarrow$ He, Ar, He + Ar, Xe + Ar, rel. to min.	100			
1.3-140G	$\mu \rightarrow$ He rel. to 1.3 GeV/c μ	101			
0.3-2.2G	$\mu \rightarrow$ Plast. Scint.; also δS	105			
0.3-70G	$\mu, \pi \rightarrow$ Ar; also δS	108			
0.16-1.8G	$\mu \rightarrow$ Ar	109			
0.07-1.5G	$\mu \rightarrow$ Air rel. to min.	113			
0.2-5G	$\mu, \pi \rightarrow$ Kr + CH_4	115			
0.4-40G	$\mu, \pi \rightarrow$ Ar + Ethylene, rel. to min.	119			
0.5-50G	$\mu \rightarrow$ Xe, Xe + He, rel. to min.	120			
0.3-7G	$\pi \rightarrow$ Emulsion, rel. to min.	121			
600M	$\pi \rightarrow$ Ne; δS only	514			
1.5-16G	$\pi, K \rightarrow$ Ar + 5% CH_4	625			
9G	$\pi \rightarrow$ Ar + 5% CH_4, 60% He + 10% CH_4 + 30% Ar, Kr + 5% CH_4	626			
5G	$\pi \rightarrow$ Emulsion rel. to min.	645			
25-150G	$\pi \rightarrow$ Ar+20% CO_2	868			
1.35G	$\pi \rightarrow$ Ge (cryst., chann., random)	874			
1.5G	μ				
1.5,40G	$\pi^- \rightarrow$ Ar+7%CH_4; δS only	897			
5-12G	$\pi^- \rightarrow$ Ar+7%CO_2	898			
3G	$\pi^- \rightarrow$ Ar+7%CO_2; δS only	916			

TABLE V
BIBLIOGRAPHY

1	G. Mezey, Z. Szökefalvi-Nagy, C. S. Badinka: Measurement of the boron distribution in ^{10}B-implanted silicon by the (n,α) nuclear reaction. Thin Solid Films *19*, 173-75 (1973).	R, δR. 10-40 keV ^{10}B $\rightarrow$ Si
2	S. K. Allison, C. S. Littlejohn: Stopping power of various gases for lithium ions of 100-450-keV energy. Phys. Rev. *104*, 959-61 (1961).	S. 100-450 keV Li $\rightarrow$ H$_2$, He, Ar, Air
3	G. Anianson: New method for measuring the α-particle range and straggling in liquids. Phys. Rev. *98*, 300-02 (1955).	R. 5.3 MeV α $\rightarrow$ H$_2$O, Ethyl alcohol, C$_6$H$_6$, pyridine, n-neptane iso-octane, 2-2-4-trimethyl pentane
4	R. K. Appleyard: The stopping power of liquid water. Proc. Camb. Phil. Soc. *47*, 443-49 (1951).	S. 4-5 MeV α $\rightarrow$ H$_2$O rel. to air
5	F. Schulz, K. Wittmaack, J. Maul: Implications in the use of secondary ion mass spectroscopy to investigate impurity concentration profiles in solids. Rad. Effects. *18*, 211-15 (1973).	R, δR. 20 keV ^{11}B $\rightarrow$ Si
6	W. K. Hofker, H. W. Werner, D. P. Oosthoek, H. A. M. de Grefte: Influence of annealing on the concentration profiles of boron implantations in silicon. Applied Physics *2*, 265-78 (1973).	R, δR. 70 keV B $\rightarrow$ Si
7	M. Bader, W. A. Wenzel, W. Whaling: Proton stopping cross sections of lithium. Phys. Rev. *92*, 1085 (1953).	S. 440 keV p $\rightarrow$ Li
8	M. Bader, R. E. Pixley, F. J. Moser, W. Whaling: Stopping cross sections of solids for protons, 50-600 keV. Phys. Rev. *103*, 32-38 (1956).	S. 50-600 keV p $\rightarrow$ Cu, Au, Pb, LiF, CaF$_2$ 75 keV - 1.4 MeV p $\rightarrow$ Li, 50 kev - 2.6 MeV p $\rightarrow$ Be, 200 - 600 keV p $\rightarrow$ Al, Mn, Ta rel. to Au; $\rightarrow$ Ca, V, Cr, Fe, Co, Ni, Cu, Zn rel. to Mn
9	S. Barile, R. Webeler, G. Allen: Determination of the ranges and straggling of low-energy alpha particles in a cloud chamber. Phys. Rev. *96*, 673-78 (1954).	R, δR. 480-615 keV α $\rightarrow$ Air
10	L. F. Bates: On the range of α-particles in rare gases. Proc. Roy. Soc. A *106*, 622-32 (1924).	R. 6.1-MeV α $\rightarrow$ He, Ne, O$_2$, Ar, Kr, Xe
11	G. Bertolini, M. Bettoni: On the range-energy relation for slow α-particles in argon. Nuovo Cimento *1*, 644-50 (1955).	R. 0.25 - 3.5 MeV α $\rightarrow$ Ar
12	F. N. Schwettmann: Enhanced diffusion during the implantation of arsenic in silicon. Appl. Phys. Lett. *22*, 570-72 (1973).	R, δR. 120 keV As $\rightarrow$ Si
13	W. K. Chu, B. L. Crowder, J. W. Mayer, J. F. Ziegler: Range distributions of implanted ions in SiO$_2$, Si$_3$N$_4$, and Al$_2$O$_3$. Appl. Phys. Letters *22*, 490-92 (1973).	R, δR. See No. 539

14	H. Bichsel, R. F. Mozley, W. A. Aron: Range of 6- to 18-MeV protons in Be, Al, Cu, Ag and Au. Phys. Rev. *105*, 1788-95 (1957).	R. 6-18 MeV p → Be, Al, Cu, Ag, Au
15	R. Shnidman, R. M. Tapphoon, K. N. Geller: Recoil fraction technique for measuring nuclear and electronic stopping powers. Appl. Phys. Letters *22*, 551-53 (1973).	S. 0.4 - 1.5 MeV C → C
16	P. M. S. Blackett, D. S. Lees: Further investigations with a Wilson chamber. II. The range and velocity of recoil atoms. Proc. Roy. Soc. *A134*, 658-71 (1932).	R. 100-800 keV α, 100-500 keV p → Air
17	W. K. Hofker, H. W. Werner, D. P. Oosthoek, H. A. M. de Grefte: Profiles of boron implantation in silicon measured by secondary ion mass spectroscopy. Rad. Effects *17*, 83-90 (1973).	R. δR. 30-75 keV B → Si
18	K. Wittmaack, F. Schulz, J. Maul: Nongaussian range profiles in amorphous solids. Phys. Letters *43A*, 477-78 (1973).	R, δR. 100 keV ^{11}B → Si (amorphous)
19	J. Maul, F. Schulz, K. Wittmaack: Determination of implantation profiles in solids by secondary ion mass spectrometry. Phys. Letters *41A*, 177-78 (1972).	R. δR. 20-22 keV ^{11}B → Si (amorphous)
20	J. K. Boggild, L. Minnhagen: A cloud-chamber study of the disintegration of lithium by slow neutrons. Phys. Rev. *75*, 782-85 (1949).	R. 880-keV p → air
21	S. Datz, C. D. Moak, B. R. Appleton, M. T. Robinson, O. S. Oen: Energy dependence of channelled ion energy loss spectra, in D. W. Palmer, M. W. Thompson, P. D. Townsend (eds); *Atomic Collison Phenomena in Solids*.North-Holland, Amsterdam, pp. 374-87 (1970).	S. 15-60 MeV ^{127}I → Au. Cryst.
22	J. L. Whitton, G. Carter: The implantation profiles of energetic heavy ions in GaAs, GaP, and Ge in D. W. Palmer, M. Thompson, P. D. Townsend (eds); *Atomic Collision Phenomena in Solids*. North-Holland, Amsterdam, pp. 615-32 (1970).	R, δR. 10-40 keV ^{35}S → GaAs. 40 keV ^{35}S → Ge 40 keV Kr, Na → GaAs, all cryst.
23	G. P. Boicourt, J. E. Brolley: Po α ranges in various counting mixtures. Rev. Sci. Instr. *25*, 95-96 (1954).	R. 5.3 MeV α → H_2, He, N_2, Ar, Ar + 1.5% CO_2, Ar + 10% CO_2, CH_4, Ar + 10% CH_4, Kr, Kr + 1% CO_2
24	W. H. Bragg, R. Kleeman: On the α particles of radium and their loss of range in passing through various atoms and molecules. Phil. Mag. *10*, 318-40 (1905).	S. 7.7 MeV α → CH_3Br, CH_3I, C_2H_5Cl, CCl_4, $C_2H_5OC_2H_5$, H_2, Al, Cu, Ag, Sn, Pt, Au, all rel. to air
25	G.H. Briggs: The decrease of velocity of α particles from radium C. Proc. Roy. Soc. *A114*, 341-54 (1927).	S. 3 - 5.3 MeV α → mica rel. to air
26	J. E. Brolley, F. L. Ribe: Energy Loss by 8.86 MeV deuterons and 4.43 MeV protons. Phys. Rev. *98*, 1112-14 (1955).	S. 4.43 MeV p → H_2, air, Kr. 8.86 MeV d → H_2, He, N_2, O_2, Ne, Ar, Kr, Xe

27	G. Dearnaley, M. A. Wilkens, P. D. Goode, J. H. Freeman, G. A. Gard: The range distribution of radioactive ions implanted into silicon crystals, in D. W. Palmer, M. W. Thompson, P. D. Townsend: *Atomic Collision Phenomena in Solids*. North-Holland, Amsterdam, p. 623-55 (1970).	R, δR. 40-120 keV ^{32}P, 10-100 keV ^{24}Na, 10-40 keV ^{35}S, ^{64}Cu, ^{85} Kr → Si, cryst. 40, 120 keV ^{32} P → Si, amorph.
28	M. Burcham: The range energy relations for protons of intermediate energy in air. Phil. Mag. *44*, 211-13 (1953).	R. 1-12 MeV p → Air
29	H. G. de Carvalho: Range of α-particles in water and ice. Phys. Rev. *78*, 330 (1950).	R. 5.3, 7.7 MeV α → H_2O. Liq., sol.
30	H. G. de Carvalho, H. Yagoda: The range of alpha-particles in water. Phys. Rev. *88*, 273-78 (1952).	R. 5.3, 7.7 MeV α → H_2O. Liq., sol.
31	T. S. la Chapelle: Range of Np^{237} alpha particles in air. AECD 2496 p. 1-16 (1947).	R. 4.7 MeV α → air
32	A. B. Chilton, J. N. Cooper, J. C. Harris: The stopping power of various elements for protons of energies from 400 to 1050 keV. Phys. Rev. *93*, 413-18 (1954).	S. 400-1050 keV p → N_2, Ne, Ar, Kr, Xe, Ni, Cu
33	R. L. Clarke, G. A. Bartholomew: Proton range-energy relation. Phys. Rev. *76*, 146-47 (1949).	R. 142, 194 keV p → D_2 + D_2O
34	F. R. Buskirk, J. N. Dyer, H. D. Hanson, R. Seng, R. H. Weidmu: Emulsion grain density in the extreme relativistic range, in *Proc. 5th Int. Conf. on Nuclear Photography*. CERN Report 65/4 No. 2 p. IX 9-13 (1965).	S. 0.2-16 GeV π^- → Emulsion. Rel. to minimum
35	M. Y. Colby, T. N. Hatfield: An apparatus for the measurement of alpha-particle range and relative stopping power of gases. Rev. Sci. Instr. *12*, 62-66 (1941).	R. 5.3 MeV α → Air, Ar
36	P. N. Cooper, V. S. Crocker, J. Walker: The relative stopping-power of hydrogen and of helium for slow α-particles. Proc. Phys. Soc. *66A*, 658-59 (1953).	R. 1.5-4.5 MeV α → H_2 He. Rel. to air
37	P. N. Cooper, V. S. Crocker, J. Walker: Range-energy data from the ^{10}B(n,α)^{7}Li and ^{6}Li(n,t)^{4}He reactions. Proc. Phys. Soc. *66A*, 660-61 (1953).	R. 1.47 MeV α, 2.75 MeV ^{3}H → air
38	A. J. Herz, B. Stiller: Relativistic increase in track blob density in various nuclear emulsions. *Proc. 5th Int. Conf. on Nuclear Photography*. CERN Report 65/4 Vol. 2, p. IX 23-31 (1965).	S. 515 MeV/C π^- → Emulsion. Rel. to electrons.
39	I. C. Cornog, W. Franzen, W. E. Stephens: Range of protons from N^{14}(n,p)C^{14}. Phys. Rev. *74*, 1-4 (1948).	R. 561 keV p → N_2
40	C. M. Creenshaw: The loss of energy of hydrogen ions traversing various gases. Phys. Rev. *62*, 54-64 (1942).	S. 60-340 keV p, d → H_2, D_2, He, H_2O rel. to air
41	C. F. Barnaby: The energy loss of singly charged heavy relativistic particles in an organic material. Proc. Phys. Soc. *77*, 1149-56 (1961).	S. 0.5-10 GeV μ^+ → plastic scintillators

42	S. Devons, J. H. Towle: Range-velocity relationship for ^{7}Li-ions in solids. Proc. Phys. Soc. *69A*, 345-47 (1956).	S. 2.74 MeV ^{7}Li → Al, Cu, Au
43	J. K. Boggild, K. H. Hansen, J. E. Hooper, M. Scharff, P. K. Aditya: The range-energy relation of nuclear emulsions studied on the ranges of secondary particles from well-known decays. Nuovo Cimento Suppl. *26*, 303-35 (1962).	R. 100, 500 MeV, π 810 MeV μ → Emulsion
44	A. Eckardt: Geschwindigkeitsverlust von H-Kanalstrahlen beim Durchgang durch feste Körper. Ann. Physik *5*, 401-28 (1930).	S. 30-50 keV p → Celluloid
45	G. G. Eichholz, N. J. Harrick: Absorption of alpha-particles in gases. Phys. Rev. *76*, 589 (1949).	R. 5.3 MeV α → Ar, D_2, CH_4
46	R. H. Ellis, Jr., H. H. Rossi, G. Failla: Stopping power of polystyrene and acetylene for alpha-particles. Phys. Rev. *86*, 562-63 (1952).	S. 5-3 MeV α → polystyrene rel. to acetylene
47	R. H. Ellis, Jr., H. H. Rossi, G. Failla: Stopping power of water films. Phys. Rev. *97*, 1043-47 (1955).	R. 5.3 MeV α → H_2O
48	G. E. Evans, P. M. Stier, C. F. Barnett: The stopping of heavy ions in gases. Phys. Rev. *90*, 825-32 (1953).	R. 20-250 keV He, N, Ne, Ar → He, N_2, Ar, Air
49	G. Alexander, R. H. W. Johnston: On the relation between blob-density and velocity of singly charged particles in G-5 emulsion. Nuovo Cimento *5*, 363-77 (1957).	S. $0.79 < \beta < 0.94$ π, μ → Emulsion
50	H. Faraggi: La réaction ^{10}B(n,α)Li* et la relation parcours-énergie pour les particules α d'énergie inférieure à 2 MeV. C. R. Acad. Sci. *229*, 1223-25 (1949).	R. 0.5-2.0 MeV α → Air
51	M. Förster: Reichweiten von α-Strahlen und chemische Bindung. Ann. Physik *27*, 373-88 (1936).	R. 5.3 MeV α → H_2, O_2, H_2O
52	S. Baskin, J. R. Winkler: Relative ionization by cosmic ray μ mesons in a liquid scintillator. Phys. Rev. *92*, 464-67 (1953).	S. 385-2200 MeV μ^+ → Xylene rel. to min.
53	Chr. Gerthsen: Über Ionisation und Reichweite von H-Kanalstrahlen in Luft und Wasserstoff. Ann. Physik *5*, 657-69 (1930).	R. 20-64 keV p → Air, H_2
54	C. W. Gilbert: The disintegration of boron by slow neutrons. II. Proc. Camb. Phil. Soc. *44*, 447-52 (1948).	R. 1.53-2.97 MeV α → Air
55	T. Bowen: Ionization energy loss of mesons in a sodium iodide scintillation crystal. Phys. Rev. *96*, 754-64 (1954).	S. 61-222 MeV π, 245-5230 MeV μ → NaI
56	G. W. Gobeli: Range-energy relation for low-energy alpha particles in Si, Ge and InSb. Phys. Rev. *103*, 275-78 (1956).	R. 0.70-4.45 MeV α → Si, Ge, InSb, Al, Cu, Ag, Au
57	G. E. Evans, C. F. Barnett, P. M. Stier, V. L. de Rito: Extrapolated ionization ranges of ions heavier than protons. ORNL-1278, 17-21 (1952).	R. 50-300 keV p, He, N, Ne, N_2, 50-160 keV Ar → He, N_2, Ar, Air

58	F. J. Congel, P. S. McNulty: Relativistic energy loss by ionization in nuclear emulsion. Phys. Rev. *176*, 1615-20 (1968).	S. 5-24 GeV/c p, 5 GeV/c $\pi \rightarrow$ Emulsion. Rel. to min.
59	D. W. Green, J. N. Cooper, J. C. Harris: Stopping cross section of metals for protons of energies from 400 to 1000 keV. Phys. Rev. *98*, 466-70 (1955).	S. 0.4-1.0 MeV p $\rightarrow$ Mn, Cu, Ge, Sn, Se, Ag, Sb, Au, Pb, Bi
60	B. Grinberg, M. Marquès da Silva: Courbes d'ionisations dans le tetrafluorore de carbone et l'hexafluorore des soufre relatives aux rayons α du polonium. J. phys. radium *6*, 69-70 (1935).	R. 5.3 MeV $\alpha \rightarrow CF_4$, SF_6 rel. to air
61	R. W. Gurney: The stopping-power of gases for alpha-particles of different velocities. Proc. Roy. Soc. *A107*, 340-49 (1925).	S. 5.3, 6.1 MeV $\alpha \rightarrow H_2$, He, O_2, Ne, Ar, Kr, Xe rel. to air
62	D. Hackman, O. Haxel: Energie und Reichweite langsamer α-Strahlen. Z. Physik *120*, 486-92 (1943).	R. 1-3 MeV $\alpha \rightarrow$ air
63	F. E. Hammer, F. E. Hoecker: A new method of measuring the stopping power of several materials for alpha-particles. Rev. Sci. Instr. *20*, 394-98 (1949).	S. 5.3 MeV $\alpha \rightarrow$ Al, mica, nylon, polystyrene rel. to air
64	G. I. Harper, I. Salaman: Measurements on the ranges of α-particles. Proc. Roy. Soc. *A127*, 175-85 (1930).	R. 5.3-7.7 MeV $\alpha \rightarrow H_2$, Ne, O_2, Air
65	T. N. Hatfield, A. E. Lockenwitz, M. Y. Colby: The relative stopping power of gases for alpha particles from polonium. J. Franklin Inst. *247*, 133-36 (1949).	S. 5.3 MeV $\alpha \rightarrow H_2$, N_2, O_2, N_2O, CO_2, H_2S, C_3H_8, C_3H_6 (cyclopropane), C_3H_6 (propylene), C_4H_{10} (n-butane), C_4H_{10} (iso-butane), C_4H_8 (butene-1), C_4H_8 (butene 2), C_4H_8 (iso-butene), CCl_4, C_2H_5Cl, CCl_2F_2, C_2H_5I.
66	L. S. Haworth, L. D. P. King: The stopping power of lithium for low energy protons. Phys. Rev. *54*, 48-50 (1938).	S. 35-400 keV p $\rightarrow$ Li
67	Z. H. Heller, D. J. Tendam: The stopping power of metals and semiconductors. Phys. Rev. *84*, 905-09 (1951).	S. 9 MeV d $\rightarrow$ Si, Ni, Cu, Ge, Zr, Rh, Ag, Sn, Air rel. to Al
68	A. Crispin, P. J. Hayman: Ionization loss of muons in a plastic scintillator. Proc. Phys. Soc. *83*, 1051-58 (1964).	S. 0.4-100 GeV/c $\mu^+ \rightarrow$ Plast. scint.
69	D. A. Eyeions, B. G. Owen, B. T. Price, J. G. Wilson: The ionization loss of relativistic μ-mesons in neon. Proc. Phys. Soc. *68*, 793-800 (1955).	S. 1.5-100 GeV/c $\mu \rightarrow$ Ne
70	E. L. Hubbard, K. R. MacKenzie: The range of 18-MeV protons in aluminum. Phys. Rev. *85*, 107-11 (1952).	R. 18 MeV p $\rightarrow$ Al
71	T. Huus, C. B. Madsen: Proton stopping power of gold. Phys. Rev. *76*, 323 (1949).	S. 364, 992 keV p $\rightarrow$ Au

72	J. R. Fleming, J. J. Lord: Ionization loss in nuclear emulsions. Phys. Rev. *92*, 511-12 (1953).	S. 24-224 MeV $\pi \rightarrow$ Emulsion. Rel. to min.
73	S. K. Ghosh, G. M. D. B. Jones, J. G. Wilson: Ionization by relativistic μ-mesons in oxygen. Proc. Phys. Soc. *67*, 331-42 (1954).	S. 0.5-30 GeV/c $\mu \rightarrow O_2$
74	S. K. Ghosh, G.M. D. B. Jones, J. G. Wilson: Ionization in oxygen by μ-mesons. Proc. Phys. Soc. *65*, 68-69 (1952).	S. 0.5-30 GeV/c $\mu \rightarrow O_2$
75	R. B. Leachman, H. Atterling: Nuclear collision stopping of astatine atoms. Ark. Fysik. *13*, 101-08 (1957).	R. 3 MeV At $\rightarrow$ Al, Au
76	D. Kahn: The energy loss of protons in metallic foils and mica. Phys. Rev. *90*, 503-09 (1953).	S. 400-1350 keV p $\rightarrow$ Be, Al, Cu, Au, mica
77	E. L. Kelly: Experimental determination of stopping powers using alpha-particles of 15-37 MeV. Phys. Rev. *75*, 1006-07 (1949).	S. 28, 37 MeV $\alpha \rightarrow$ Cu, Ag, Ta, Bi, Th rel. to Al
78	A. B. Lillie: The disintegration of oxygen and nitrogen by 14.1 MeV neutrons. Phys. Rev. *87*, 716-22 (1952).	R. 0.5-5 MeV ^{13}C, 0.5-7 MeV ^{11}B $\rightarrow$ Air
79	R. L. Sen Gupta: Specific ionization of cosmic ray particles. Proc. Natl. Inst. Sci. India *9*, 295-300 (1943).	S. rel. to min. 0.2-20 GeV/c $\mu, \pi \rightarrow$ Air
80	G. Hall: Ionization loss by cosmic-ray mu-mesons in argon. Can. J. Phys. *37*, 189-202 (1959).	S, δS. $3.1 < P/mc < 6.2$ μ^+ $\rightarrow$ Ar
81	G. J. van der Maas, J. I. Yntema: A determination of the mean range of the alpha particles of ^{235}U. Physica *15*, 807-24 (1949).	R. 4.4 MeV α rel. to 4.2 MeV $\alpha \rightarrow$ Air
82	M. Mäder: Reichweite und Gesamt-ionisation von α-Strahlen in Gasen. Z. Physik *77*, 601-15 (1932).	R. 5.3 MeV $\alpha \rightarrow$ Air, CO_2, N_2, O_2
83	C. B. Madsen, P. Venkateswarlu: Proton stopping power of solid beryllium. Phys. Rev. *74*, 648-49 (1948).	S. 500-1500 keV p $\rightarrow$ Be
84	C. B. Madsen: Proton stopping power and energy straggling of protons. Kgl. Danske Videnskab. Selskab Mat. Fys. Medd. *27*, No. 13, 1-21 (1953).	S. δS. 350-2000 keV p $\rightarrow$ Be, Al, Cu, Ag, mica
85	G. Mano: Recherches sur l'absorption des rayons α. Ann. de Physique *1*, 408-531 (1934).	S. 4.2-7.7 MeV $\alpha \rightarrow H_2$, He, Ne, Ar, Air
86	D. G. Jones, R. H. West, A. W. Wolfendale: The rate of energy loss of cosmic ray muons in a neon-methane mixture. Proc. Phys. Soc. *81*, 1137-39 (1963).	S. 0.3-30 GeV/c $\mu \rightarrow$ Ne + CH_4
87	E. Marsden, H. Richardson: The retardation of α particles by metals. Phil. Mag. *25*, 184-93 (1913).	R. 4-8 MeV $\alpha \rightarrow$ Al, Cu, Ag, Sn, Pt, Au, mica rel. to air
88	E. Marsden, T. S. Taylor: The decrease in velocity of α-particles in passing through matter. Proc. Roy. Soc. *A88*, 443-54 (1913).	S. 5-8 MeV $\alpha \rightarrow$ Al, Cu, Au, air, mica

89	M. McInally: The differential stopping power of liquid water for α-particles. Proc. Roy. Soc. *A237*, 28-38 (1956).	R. 4-6 MeV $\alpha \rightarrow H_2O$ rel. to air
90	L. Meitner, K. Freitag: Über die α-Strahlen des ThC+C' und ihre Verhalten beim Durchgaug durch verschiedene Gase. Z. Physik *37*, 481-517 (1926).	R. 5.3 MeV $\alpha \rightarrow$ Air, H_2, H_2O, CH_4, N_2, O_2, CO, CO_2, SO, SO_2, CH_3Br
91	W. Michl: Über die Reichweite der α-Strahlen in Flüssigkeiten. Sitzber. Akad. Wiss. Wien Mathematich-naturwissenschaftliche Klasse *123*, 1965-99 (1914).	R. 5.3 MeV $\alpha \rightarrow CS_2$, H_2O, C_2H_5OH, ether, benzene, anilin, glycerin, chloroform
92	R. G. Mills: A cloud chamber investigation of low energy range-energy relations. UCRL-1815, 1-89 (1952).	R. 50-250 keV p, 100-360 keV α, 30-110 keV O $\rightarrow$ He, O_2, H_2O
93	R. Naidu: Sur les courbes d'ionisation des rayons α du polonium dans les gaz rares. J. Phys. Radium *5*, 575-77 (1934).	S. 3-7 MeV $\alpha \rightarrow$ He, Ne, Ar, Kr, Xe. All rel. to air
94	B. Jongejans: On the grain density in Ilford G-5 emulsion of singly charged relativistic particles. Nuovo Cimento *6*, 623-43 (1960).	S. 140-170 GeV π, $\mu \rightarrow$ emulsion rel. to 5.6 GeV $\pi \rightarrow$ emulsion
95	D. B. Parkinson, R. G. Herb, J. L. Bellamy, C. M. Hudson: The range of protons in aluminum and air. Phys. Rev. *52*, 75-79 (1937).	R. 0.1-2.0 MeV p $\rightarrow$ air, Al
96	J. Phelps, W. F. Huebner, F. Hutchinson: The stopping power of organic foils for 6 MeV alphas. Phys. Rev. *95*, 441-44 (1954).	S. 6 MeV $\alpha \rightarrow CH_2$, plastics, Al rel. to air
97	H. R. von Traubenberg, K. Phillipp: Über zwei direkte Methoden zur Bestimmung der Reichweite von α-Strahlen in Flüssigkeiten und Gasen. Z. Physik *5*, 404-09 (1921).	R. 5.5 MeV $\alpha \rightarrow H_2O$, CO, CO_2, CH_3Br, CH_3I, Cl_2, HCl, NH_3, rel. to air
98	K. Phillip: Bremsung der α-Strahlen in Flüssigkeiten und Dämpfen. Z. Physik *17*, 23-41 (1923).	R. 5.5 MeV $\alpha \rightarrow$ Air, CO_2, H_2O, C_2H_5OH, C_6H_6, C_5H_5N
99	J. A. Phillips: The energy loss of low energy protons in some gases. Phys. Rev. *90*, 532-37 (1953).	S. 10-80 keV p $\rightarrow H_2$, He, N_2, O_2, Ar, Kr, H_2O, CO_2, CCl_4
100	R. G. Kepler, C. A. d'Andlau, W. B. Fretler, L. F. Hansen: Relativistic increase of energy loss by ionization in gases. Nuovo Cimento *7*, 71-86 (1958).	S. 0.3-12 GeV $\mu \rightarrow$ He, Ar, He + Ar, rel. to min.
101	R. E. Lanou, H. L. Kraybill: Ionization loss by μ-mesons in helium. Phys. Rev. *113*, 657-61 (1959).	S. 1.3-140 GeV/c $\mu \rightarrow$ He rel. to 1.3 GeV/c
102	W. Reusse: Energieverluste langsamer Kanalstrahlen beim Durchgang durch feste Körper. Ann. Physik *15*, 256-58 (1932).	S. 30-55 keV p $\rightarrow$ Celluloid
103	H. K. Reynolds, D. N. F. Dunbar, W. A. Wenzel, W. Whaling: The stopping cross section of gases for protons, 30-600 keV. Phys. Rev. *92*, 742-48 (1953).	S. 30-600 keV p $\rightarrow H_2$, He, O_2, air, N_2, Ne, Ar, Kr, Xe, hydrocarbons

104	H. L. Reynolds, N. F. Scott, A. Zucker: Range and charge of energetic nitrogen ions in nickel. Phys. Rev. *95*, 671-74 (1954).	R. 8-29 MeV N → Ni
105	C. H. Millar, E. P. Hincks, G. C. Hanna: A large-area liquid scintillation counter and some measurements on high-energy cosmic-ray particles. Can. J. Phys. *36*, 54-72 (1958).	S, δS. 0.3-0.8 GeV p, 0.3-2.2 GeV μ → liquid scintillators
106	W. Riezler: Bremsvermögen von Glimmer für Alphateilchen kleiner Reichweite. Ann. Physik *35*, 350-53 (1939).	S. 1-4 MeV α → mica
107	W. Riezler: Ionisierung und Reichweite von Alphateilchen in Schwefelhexafluorid. Ann. Physik *35*, 354-58 (1939).	R. 0.5-4 MeV α → SF_6
108	J. K. Parry, H. D. Rathgeber, J. L. Rouse: Ionization of cosmic ray mesons in argon. Proc. Phys. Soc. *66*, 541-48 (1953).	S, δS. 0.3-70 GeV/c μ, π → Ar
109	J. E. Kupperian, Jr., E. P. Palmatier: The ionization loss of energy of relativistic mu-mesons in argon. Phys. Rev. *85*, 1043 (1952).	S. 160-1800 MeV μ → Ar
110	S. Rosenblum: Recherches expérimentales sur le passage des rayons α a travers la matière. Ann. de Physique *10*, 408-71 (1928).	S. 5.3 - 7.7 MeV α → Li, Al, Fe, Ni, Cu, Zn, Mo, Pd, Ag, Cd, Sn, Pt, Au, Pb, mica, AuAg alloys, Ag-Cu alloys
111	B. V. Rybakov: Ranges of protons in medium and heavy elements. Zh. Exp. Teor. Fiz. *28*, 651-54 (1955) [Engl. trans. Sov. Phys. JETP *1*, 435-38 (1955)].	R. 1-7 MeV p → Fe, Cu, Mo, Cd, Sn, Pd, Ta rel. to Al
112	K. Schmieder: Bremsvermögen und Trägerbildung der α-Strahlen in Gasen. Ann. Physik *35*, 445-64 (1939).	R. 5.3 MeV α → Air, N_2, O_2, NO, N_2O, NO_2, CO, CO_2, CH_4, C_2H_4, Ne, Ar, Kr, N_2+O_2, N_2+2O_2, $2N_2+O_2$
113	R. S. Carter, W. L. Whittemore: The relative specific ionization of fast mesons. Phys. Rev. *87*, 494-99 (1952).	S. 70-1500 MeV/c μ → air
114	D. H. Simmons: The range-energy relation for protons in aluminum. Proc. Phys. Soc. *65A*, 454-56 (1952).	R. 10 MeV p → Al
115	J. Becker, P. Chanson, E. Nageotte, P. Treille, B. T. Price, P. Rothwell: The increase of ionization with momentum for energetic cosmic-ray particles. Proc. Phys. Soc. *65*, 437-48 (1952).	S. 0.2-5 GeV/c μ,π → Kr + CH_4.
116	C. P. Sonett, K. R. MacKenzie: Relative stopping power of various metals for 20 MeV protons. Phys. Rev. *100*, 734-32 (1955).	S. 20.6 MeV p → Ni, Cu, Nb, Pd, Ag, Cd, In, Ta, Pt, Au, Th, rel. to Al.
117	T. S. Taylor: On the retardation of alpha rays by metals and gases. Phil. Mag. *18*, 604-19 (1909).	S. 7.7 MeV α → Au, Sn, Pb, Al, H_2, paper, collodium, rel. to air
118	T. S. Taylor: The range and ionization of the alpha particles in simple gases. Phil. Mag. *26*, 402-10 (1913).	R. 5.3, 5.5 MeV α → H_2, He, O_2, air

119	P. Goodman, K. P. Nicholson, H. D. Rathgeber: The ionization of cosmic-ray particles. Proc. Phys. Soc. *64*, 6-7 (1951).	S. rel. to min. 0.4-40 GeV/c μ, π → Ar + ethylene
120	A. Rousset, A. Lagarrique, P. Musset, P. Rançon, X. Santéron: Relativistic increase of ionization in xenon. Nuovo Cimento *14*, 365-75 (1959).	S. 0.5-50 GeV/c μ → Xe, Xe + He rel. to min.
121	B. Stiller, M. M. Shapiro: Ionization loss at relativistic velocities in nuclear emulsion. Phys. Rev. *92*, 735-41 (1953).	S. 1-4 GeV/c p, 0.3-7 GeV/c π → emulsion rel. to min.
122	J. G. Teasdale: Stopping of various elements relative to aluminum for 12 MeV protons. NP 1368, 1-16 (1949).	S. 12 MeV p → Ni, Cu, Rh, Pd, Ag, Cd, In, Ta, Pt, Au, Th
123	H. R. von Traubenberg: Über eine Methode zur direkten Bestimmung der Reichweite von α-Strahlen in festen Körpern. Z. Physik *2*, 268-76 (1920).	R. 7.7 MeV α → H_2, He, Li, O_2, Mg, Al, Ca, Fe, Ni, Au, Zn, Ag, Cd, Sn, Pt, Cu, Tl, Pb
124	W. K. Chu, J. F. Ziegler, I. V. Mitchell, W. D. Mackintosh: Energy-loss measurements of ^{4}He ions in heavy metals. Appl. Phys. Letters *22*, 437-39 (1973).	S. 2.0 MeV α → Al, Si, V, Fe, Co, Ni, Cu, In, Ge, Mo, Sb, Te, Gd, Hf, Ta, W, Ir, Pt, Au, Pb
125	S. Furukawa, H. Matsumura: Backscattering study on lateral spread of implanted ions. Appl. Phys. Letters *22*, 97-98 (1973).	R, δR, $\delta R_\perp$. 180 keV Kr → Si
126	P. Cüer, J. P. Lonchamp: Étude de la réaction des neutrons thermiques sur le bore. C. R. Acad. Sci. *232*, 1824-26 (1951).	R. 1.47, 1.78 MeV α, 0.84, 1.02 MeV ^{7}Li → Air, Emulsion
127	G. J. Clark, G. Dearnaley, D. V. Morgan, J. M. Poate: The stopping power of protons channelled through CsI crystals. Phys. Letters *30A*, 11-12 (1969).	S. 4 MeV p → CsI (cryst.)
128	G. Schwartz, M. Trapp, R. Schimko, G. Butzke, K. Rogge: Concentration profiles of implanted boron ions in silicon from measurements with the ion microprobe. Phys. Stat. Sol. (a) *17*, 653-58 (1973).	R, δR. 50-300 keV B → Si
129	S. D. Warshaw: The stopping power of protons in several metals. Phys. Rev. *76*, 1759-65 (1949).	S. 50-400 keV p → Be, Al, Cu, Ag, Au
130	W. A. Wenzel, W. Whaling: The stopping cross section of D_2O ice. Phys. Rev. *87*, 499-503 (1952).	S. 18-540 keV p → D_2O (ice)
131	P. K. Weyl: The energy loss of hydrogen, helium, nitrogen and neon ions in gases. Phys. Rev. *91*, 289-96 (1953).	S. 150-450 keV p, d, α, N, Ne → H_2, He, Air, Ar
132	G. della Mea, A. V. Drigo, S. lo Russo, P. Mazzoldi, G. G. Bentini: Energy loss of H, D, and ^{4}He ions channeled through thin single crystals of silicon. Phys. Rev. Letters *27*, 1194-96 (1971).	S. 0.9-5.0 MeV p, d, α → Si cryst. and amorph.

133	H. W. Wilcox: Experimental determination of rate of energy loss for slow H^1, H^2, He^4, Li^6 nuclei in Au and Al. Phys. Rev. *74*, 1743-54 (1948).	S. 30-400 keV p, 30-650 keV d, 30-1400 keV α, 750-850 keV ^{6}Li $\rightarrow$ Al, Au
134	A. Perez, P. Thevenard, J. Davenas, C. H. S. Dupuy: F-centre profiles associated with electronic stopping power in LiF bombarded with high energy ions. Phys. Stat. Sol. (a) *18*, 189-95 (1973).	S. 56 MeV α, 28 MeV d $\rightarrow$ LiF. Dist. along path
135	D. H. Wilkinson: The stopping power of polythene and fast neutron flux measurements. Proc. Camb. Phil. Soc. *44*, 114-23 (1948).	S. 5.4 MeV α $\rightarrow$ $(CH_2)_n$. Solid rel. to gas
136	R. R. Wilson: Range and ionization measurements on high speed protons. Phys. Rev. *60*, 749-53 (1941).	S. 4 MeV p $\rightarrow$ Al, Cu, Fe, Mo, Ni, Pt, Ta, Zn rel. to air.
137	L. D. Wyly, V. L. Sailor, D. G. Ott: Protons from the bombardment of He^3 by deuterons. Phys. Rev. *76*, 1532-33 (1949).	R. 16-20.5 MeV p $\rightarrow$ air
138	S. A. Wytzes and G. J. van der Maas: A new determination of the mean ranges of the α-particle from U I and U II. Physica *13*, 49-61 (1947).	R. 5.3 MeV α $\rightarrow$ air rel. to 4.2 MeV α $\rightarrow$ air
139	H. Bichsel: Experimental range of protons in Al. Phys. Rev. *112*, 1089-91 (1958).	R. 1-6 MeV p $\rightarrow$ Al
140	R. L. Hines: Ranges of 5- to 27-keV deuterons in aluminum, copper and gold. Phys. Rev. *132*, 701-06 (1963).	R. 5-27 keV D $\rightarrow$ Al, Cu, Au
141	L. C. Northcliffe: Energy loss and effective charge of heavy ions in aluminum. Phys. Rev. *120*, 1744-57 (1960).	S. 4-41 MeV α, 10-102 MeV ^{10}B, 11-112 MeV ^{11}B, 12-123 MeV ^{12}C, 14-140 MeV ^{14}N, 16-165 MeV ^{16}O, 20-195 MeV ^{19}F, 20-200 MeV ^{20}Ne $\rightarrow$ Al
142	R. L. Wolke, W. N. Bishop, E. Eichler, N. R. Johnson, G. D. O'Kelley: Ranges and stopping cross sections of low-energy tritons. Phys. Rev. *129*, 2591-96 (1963).	R, S. 0.2-2.73 MeV t $\rightarrow$ N_2, Al, Ar, Ni, Kr, Xe
143	M. McCargo, F. Brown, J. A. Davies: A reinvestigation of the range of Na^{24} ions of keV energies in aluminum. Can. J. Chem. *41*, 2309-13 (1963).	R, δR. 1-100 keV ^{24}Na $\rightarrow$ Al
144	E. W. Valyocsik: Range and range straggling of heavy recoil atoms. UCRL 8855 (1959).	R, δR. 96.8 keV ^{224}Ra $\rightarrow$ N_2, H_2, D_2, He, Ne, Ar, 725 keV ^{226}Th $\rightarrow$ D_2, He, N_2, Ar. 140-280 keV ^{208}Po $\rightarrow$ ^{209}Bi
145	B. G. Harvey, P. F. Donovan, J. R. Morton: Range energy relation for heavy atoms. UCRL 8618, 17-20 (1959).	R. 725 keV ^{226}Th $\rightarrow$ H_2, D_2, He, N_2, Ar
146	G. H. Nakano, K. R. MacKenzie, H. Bichsel: Relative stopping power of some metallic elements for 28.7 MeV protons. Phys. Rev. *132*, 291-93 (1963).	S. rel. to Al. 28.7 MeV p $\rightarrow$ Be, Ti, V, Co, Ni, Cu, Ag, Ta, W, Ir, Au

147	H. J. Thompson: Effect of chemical structure on stopping powers for high-energy protons. UCRL 1910 (1952).	S. rel. to Cu. 270 MeV p → H_2, C, N_2, O_2, Cl_2
148	F. W. Martin, L. C. Northcliffe: Energy loss and effective charge of He, C, and Ar ions below 10 MeV/amu in gases. Phys. Rev. *128*, 1166-74 (1962).	S. 4-40 MeV He, 12-120 MeV C, 40-400 MeV Ar → H_2, N_2, Ar 12-120 MeV C → He
149	V. C. Burkig, K. R. MacKenzie: Stopping power of some metallic elements for 19.8 MeV protons. Phys. Rev. *106*, 848-51 (1957).	S. rel. to Al. 19.8 MeV p → Be, Ca, Ti, V, Fe, Ni, Cu, Zn, Nb, Mo, Rh, Pd, Ag, Cd, In, Sn, Ta, W, Ir, Pt, Au, Pb, Th
150	C. O. Hower, A. W. Fairhall: Ranges of Be^9 ions in gold and aluminum. Phys. Rev. *128*, 1163-65 (1962).	R. 2-29 MeV ^{9}Be → Al, Au
151	L. P. Nielsen: Energy loss and straggling of protons and deuterons. Kgl. Danske Videnskab. Selskab Mat. Fys. Medd. *33*, No. 6, 1-20 (1961).	S, δS. 1.5-4.5 MeV p, d → Al, Ni, Cu, Ag, Au; 1.5-4.5 MeV p → Be
152	W. R. Phillips, F. H. Read: The ranges of nitrogen ions in gold. Proc. Phys. Soc. *81*, 1-8 (1963).	R, δR. 0.4-6.4 MeV ^{15}N → Au
153	A. M. Poscanzer: Range of 1-3 MeV Ne^{22} ions in Al and the analysis of some Na^{24} recoil data. Phys. Rev. *129*, 385-87 (1963).	R. 1-3 MeV ^{22}Ne → Al
154	I. Bergström, J. A. Davies, B. Domeij, J. Uhler: The range of Rn^{222} ions of keV energies in aluminum and tungsten. Ark. Fysik *24*, 389-98 (1963).	R, δR. 2-450 keV ^{222}Rn → Al, W
155	F. Brown, J. A. Davies: The effect of energy and integrated flux on the retention of inert gas ions injected at keV energies in metals. Can. J. Phys. *41*, 844-57 (1963).	R, δR. 40 keV Ar, 30, 40, keV Ar, Xe → Al
156	L. Bryde, N. O. Lassen, N. O. Roy Poulsen: Ranges of recoil ions from α-reactions. Kgl. Danske Videnskab. Selskab Mat. Fys. Medd. *33*, No. 8, 1-28 (1962).	R, δR. 0.6-1.2 MeV ^{66}Ga → H_2, D_2, He, Ar, N_2 1.0-1.3 MeV ^{66}Ga → Cu, 1.0-1.7 MeV ^{43}K → Ar, 3.9 MeV ^{18}F → N_2
157	J. A. Davies, F. Brown, M. McCargo: Range of Xe^{133} and Ar^{41} ions of kiloelectron volt energies in aluminum. Can J. Phys. *41*, 829-43 (1963).	R, δR. 0.7 keV - 2.25 MeV Ar, 0.5 - 240 keV Xe → Al
158	J. A. Davies, B. Domeij, J. Uhler: The range of Kr^{85} ions in aluminum and tungsten in the energy interval 2-600 keV. Arkiv för Fysik *24*, 377-88 (1963).	R, δR. 2-600 keV ^{85}Kr → Al, W (both cryst.)
159	Yu. V. Gott, V. G. Telkovskiy: Energy losses of light ions in thin metallic foils. Radioteknika i Elektronika *7*, 1956-61 (1962) [Engl. trans: Rad. Eng. and Electron Phys. *7*, 1813-19 (1962)].	S. 2-15 keV H, D, He → Al, Ti, Cu, Ge, Ag, Sn, Au
160	R. L. Graham, F. Brown, J. A. Davies, J. P. S. Pringle: A new method for measuring the depths of embedded radiotracer atoms using a precision β-ray spectrometer. Can J. Phys. *41*, 1686-1701 (1963).	R. 1-40 keV ^{125}Xe → Al, W, Au

161	H. Lutz, R. Sizmann: Super ranges of fast ions in copper single crystals. Phys. Letters *5*, 113-14 (1963).	R, δR. 10-150 keV ^{85}Kr → Cu (cryst.)
162	M. McCargo, J. A. Davies, F. Brown: Range of Xe^{133} and Ar^{41} ions of keV energies in tungsten. Can J. Phys. *41*, 1231-44 (1963).	R, δR. 2-200 keV ^{133}Xe, ^{41}Ar → W, 40 keV ^{85}Kr → WO$_3$
163	G. R. Piercy, F. Brown, J. A. Davies, M. McCargo: Experimental evidence for the increase of heavy ion ranges by channeling in crystalline structures. Phys. Rev. Letters *10*, 399-400 (1963).	R, δR. 40 keV ^{85}Kr → Al (cryst.), Al$_2$O$_3$
164	D. Powers, W. Whaling: Range of heavy ions in solids. Phys. Rev. *126*, 61-69 (1962).	R. 50-500 keV N, Ne, Ar, Kr, Xe → Be, B, C, Al
165	J. Uhler, B. Domeij, S. Borg: Range distributions of W^{187} ions of keV energies in tungsten. Ark. Fysik *24*, 413-19 (1963).	R, δR. 1.6-127 keV ^{187}W → W
166	J. H. Ormrod, H. E. Duckworth: Stopping cross sections in carbon for low-energy atoms with Z $\leq$ 12. Can. J. Phys. *41*, 1424-42 (1963).	S. 10-67 keV H, 10-80 keV He, 20-70 keV ^{6}Li, ^{7}Li, 12-130 keV ^{9}Be 12-140 keV ^{11}B, ^{12}C, 15-150 keV ^{14}N, 20-140 keV ^{16}O, ^{19}F, ^{20}Ne, 20-70 keV ^{27}Na, 20-130 keV ^{24}Mg → C
167	S. Barkan: Stopping power of C for ^{210}Po α-particles. Nuovo Cimento *20*, 443-49 (1961).	R. 5.3 MeV α → C
168	B. Domeij, M. McCargo, J. A. Davies, F. Brown: Ranges of heavy ions in amorphous oxides. Bull. Am. Phys. Soc. *9*, 109 (1964).	R, δR. 5-160 keV Na, Kr, Xe → Al$_2$O$_3$
169	J. A. Davies, G. C. Ball, F. Brown: Crystallographic dependence of Xe$^+$ penetration into silicon at low energies. Bull. Am. Phys. Soc. *9*, 109 (1964).	R, δR. 20-80 keV Xe → Si (cryst.)
170	C. D. Moak, M. D. Brown: Some stopping powers for iodine ions. Phys. Rev. Letters *11*, 284-85 (1964).	S. 25-115 MeV ^{127}I → C, Al, Ni, Au
171	G. Dearnaley: The channeling of ions through silicon detectors. IEEE Trans. Nucl. Sci. *NS11*, 249-53 (1964).	S, δS. 2 MeV p → Si (cryst.)
172	B. Domeij, F. Brown, J. A. Davies, G. R. Piercy, E. V. Kornelsen: Anomalous penetration of heavy ions of keV energies in monocrystalline tungsten. Phys. Rev. Letters *12*, 363-66 (1964).	R, δR. 1-40 keV ^{125}Xe → W (cryst.)
173	J. A. Davies, G. C. Ball, F. Brown, B. Domeij: Range of energetic Xe^{125} ions in monocrystalline silicon. Can. J. Phys. *42*, 1070-81 (1964).	R, δR. 5-80 keV ^{125}Xe → Si (cryst.)
174	G. R. Piercy, M. McCargo, F. Brown, J. A. Davies: Experimental evidence for the channeling of heavy ions in monocrystalline aluminum. Can. J. Phys. *42*, 1116-35 (1964).	R, δR. 20-160 keV ^{24}Na, ^{85}Kr, ^{86}Rb, ^{125}Xe → Al (cryst.)

175	J. T. Park, E. J. Zimmermann: Stopping cross section of some hydrocarbon gases for 40-250 keV protons and helium ions. Phys. Rev. *131*, 1611-18 (1963).	S. 40-250 keV H → Air, He, CH_4, C_2H_2, C_2H_4, C_3H_8, $(CH_3)_2$, C_3H_6; 40-200 keV He → C_2H_4, C_3H_8
176	W. Meckbach, S. K. Allison: Ratio of effective charge of He beams traversing gaseous and metallic conductors. Phys. Rev. *132*, 294-304 (1963).	S. 148-920 keV He, 37-230 keV H → CD (gas. and sol. phase)
177	J. Cuevas, M. Garcia-Munoz, P. Torres, S. K. Allison: Partial atomic and ionic stopping powers of gaseous hydrogen for helium and hydrogen beams. Phys. Rev. *135*, A335-45 (1964).	S. 40-460 keV He → H_2
178	N. T. Porile: Ranges of Low-energy gallium atoms in copper and zinc. Phys. Rev. *135*, A1115-18 (1964).	R. 70-1000 keV Ga → Cu, Zn
179	J. A. Davies, J. D. McIntyre, G. A. Sims: Isotope effect in heavy ion range studies. Can. J. Chem. *39*, 611-15 (1961).	R, δR. 24 keV ^{22}Na, ^{24}Na → Al
180	J. A. Davies, G. A. Sims: The ranges of Na^{24} ions of kiloelectron volt energies in aluminum. Can J. Chem. *39*, 601-10 (1961).	R, δR. 0.7-60 keV ^{24}Na, 30 keV ^{42}K, ^{86}Rb → Al
181	J. A. Davies, J. D. McIntyre, R. L. Cushing, M. Lounsbury: The ranges of alkali metal ions of kiloelectron energies in aluminum. Can. J. Chem. *38*, 1535-46 (1960).	R, δR. 2-50 keV ^{137}Cs, 30 keV ^{24}Na, ^{86}Rb → Al
182	B. Domeij, F. Brown, J. A. Davies, M. McCargo: Ranges of heavy ions in amorphous oxides. Can. J. Phys. *42*, 1624-34 (1964).	R, δR. 0.5-160 keV ^{24}Na, ^{41}Ar, ^{85}Kr, ^{125}Xe → Al_2O_3, WO_3
183	E. V. Kornelsen, F. Brown, J. A. Davies, B. Domeij, G. R. Piercy: Penetration of heavy ions of keV energies into monocrystalline tungsten. Phys. Rev. *136*, A849-58 (1964).	R, δR. 0.3-160 keV ^{24}Na, ^{41}Ar, ^{85}Kr, ^{125}Xe, ^{138}Xe → W
184	H. Lutz, R. Sizmann: Bestimmung der Reichweite schneller schwerer Ionen in Festkörpern. Z. Naturforschg. *19a*, 1079-89 (1964).	R, δR. 25-125 keV ^{85}Kr → Cu
185	J. Gilat, J. M. Alexander: Stopping of dysprosium ions in gases and Al. Phys. Rev. *136*, B1298-1305 (1964).	R. 6-21 MeV Dy → He, N_2, Ne, Ar, Kr, Xe
186	A. R. Sattler, M. G. Silbert: Ionization of energetic silicon atoms within a silicon lattice. Bull. Am. Phys. Soc. *9*, 655-56 (1964).	S. 0.2-3.1 MeV Si → Si
187	D. A. Channing, J. L. Whitton: Effect of temperature on the channeling of Xe^{133} ions in gold. Phys. Letters *13*, 27-28 (1964).	R, δR. 40 keV Xe → Au (cryst.)
188	C. Pöhlan, H. Lutz, R. Sizmann: Überreichweiten schneller Ionen in Diamantstrukturen. Z. angew. Phys. *17*, 404-06 (1964).	R, δR. 80 keV Kr → GaAs (cryst.)
189	C. Erginsoy, H. E. Wegner, W. M. Gibson: Anisotropic energy loss of light particles of MeV energies in thin silicon single crystals. Phys. Rev. Letters *13*, 530-34 (1964).	S, δS. 2.8 MeV p → Si (cryst.)

190	A. L. Morsell: Proton energy-loss distributions from thin carbon films. Phys. Rev. *135*, A1436-43 (1964).	S, δS. 990 keV p $\rightarrow$ C
191	L. Morbitzer, A. Scharmann: Messung der Eindringtiefe von Elektronen und Ionen in dünnen Aufdampfschichten. Z. Physik *181*, 67-86 (1964).	R. 1-10 keV H, 1-12 keV He, 1-30 keV Ne, Ar $\rightarrow$ LiF, NaF, MgF_2, CaF_2, ZnS.
192	L. Morbitzer, A. Scharmann: Messung der Eindringtiefe von Heliumionen und Elektronen bis 10 keV in LiF - Aufdampfs- chichten. Z. Physik *177*, 174-78 (1964).	R. 1-10 keV H^+ $\rightarrow$ LiF
193	J. R. Young: Penetration of electrons and ions in aluminum. J. Appl. Phys. *27*, 1-4 (1956).	R. 1-25 keV H, H_2, He $\rightarrow$ Al
194	L. I. Pivovar, L. I. Nikolaichuk, V. M. Rashkovan: Passage of lithium ions through condensed targets. Zh. Eks. Teor. Fiz. *47*, 1221-27 (1964) [Engl. trans. Sov. Phys. JETP *20*, 225-29 (1965)].	S. 20-145 keV Li $\rightarrow$ C
195	W. Booth, I. S. Grant: The energy loss of oxygen and chlo- rine ions in solids. Nuclear Physics *63*, 481-95 (1965).	S. 2-24 MeV O, 4-40 MeV Cl $\rightarrow$ C, Al, Ni, Ag, Au
196	A. Buhler, T. Massam, T. Muller, A. Zichichi: Range meas- urements for muons in the GeV region. Nuovo Cimento *35*, 759-67 (1965).	R. GeV μ $\rightarrow$ Pb
197	M. M. Bredov, N. M. Okuneva: On the penetration of ions of medium energy in matter (in Russian). Dokl. Akad. Nank. SSSR *113*, 795-96 (1957).	R. 4 keV Cs $\rightarrow$ Ge
198	M. M. Bredov, J. G. Lang, N. M. Okuneva: On the problem of penetration of medium-energy ions into matter. Zh. Tech. Fiz. *28*, 252-53 (1958) [Engl. trans. Sov. Phys. Tech. Phys. *3*, 228-29, (1958)].	R. 4 keV Cs $\rightarrow$ Ge
199	R. B. Burtt, J. S. Colligon, J. H. Lech: Sorption and replace- ment of ionized noble gases at a tungsten surface. Brit. J. Appl. Phys. *12*, 396-400 (1961).	R. 2.7 keV Ar, Kr $\rightarrow$ W
200	M. I. Guseva, E. V. Inopin, S. P. Tsytko: Depth of penetra- tion and character of distribution of atoms injected into Si^{30} isotope targets. Zh. Eks. Teor. Fiz. *36*, 3-9 (1959) [Engl. trans. Sov. Phys. JETP *36*, 1-5 (1959)].	R, δR. 10-28 keV ^{30}Si $\rightarrow$ Cu, 25 keV ^{30}Si $\rightarrow$ Ta
201	B. Perovic, T. Jokic: The measurement of ranges and depth distribution of ions in the kiloelectron volt energy region in metals by means of the radioactive tracer technique. *Proc. VI*[ieme] *I n t . C o n f . P h é n o m è n e s d ' I o n i z a tion dans les Gaz*, Paris Vol. II, p. 15-19 (1963).	R, δR. 5-30 keV Xe $\rightarrow$ Ni, Mo, Cu
202	G. C. Ball, F. Brown: The range of ^{133}Xe and ^{134}Cs in tung- sten single crystals and tungsten oxide at energies of 40 and 125 keV. Can. J. Phys. *43*, 676-80 (1965).	R, δR. ^{133}Xe, ^{134}Cs 40, 125 keV $\rightarrow$ W, WO_3

203	J. H. Ormrod, J. R. MacDonald, H. E. Duckworth: Some low-energy atomic stopping cross sections. Can. J. Phys. *43*, 275-84 (1965).	10-70 keV H, 12-50 keV D, 15-65 keV He, 20-70 keV Li, 15-150 keV B, C, 20-130 keV N, O, F, Ne, 25-70 keV Na → Al; 25-80 keV Al, 20-130 keV ^{29}Si, 25-130 keV P, S, 20-125 keV Cl, 25-210 keV A, 25-65 keV K → C
204	J. M. Alexander, J. Gilat, D. H. Sisson: Neutron and photon emission from dysprosium and terbium compound nuclei: Range distributions. Phys. Rev. *136*, B1289-97 (1964).	R. 6-21 MeV Dy, Tb → H_2, D_2, Al
205	H. H. Andersen: A low-temperature technique for measurement of heavy-particle stopping powers of metals. Danish AEC Risö. Report No. 93, 1-60 (1965).	S. 5-12 MeV p, d → Al
206	A. K. Roecklein, M. V. Nakai, E. T. Arakawa, J. L. Stanford, R. D. Birkhoff: A new method for measuring range of low energy charged particles in conducting solids. ORNL 3702, 1-49 (1965).	R. 2-9 keV d → Al
207	W. H. Barkas: The range correction for electron pick-up. Phys. Rev. *89*, 1019-22 (1953).	R. 4-22 MeV ^{8}Li, 8-16 MeV ^{8}B → Emulsion
208	See ref. no. 148.	
209	R. Mather, E. Segrè: Range-energy relation for 340 MeV protons. Phys. Rev. *84*, 191-93 (1951).	R. 340 MeV p → Be, C, Al, Cu, Sn, Pb
210	H. G. Clerc, H. Wäffler, F. Berthold: Reichweite von Li^8-Ionen der Energie 40-450 keV in Wasserstoff, Deuterium und Helium. Z. Naturforschg. *16a*, 149-54 (1961).	R. 40-450 keV ^{8}Li → H_2, D_2, He
211	W. H. Webb, H. L. Reynolds, A. Zucker: Nitrogen-induced nuclear reactions in aluminum. Phys. Rev. *102*, 149-52 (1956).	R. 4-28 MeV N → Al
212	H. L. Reynolds, A. Zucker: Range of nitrogen ions in emulsion. Phys. Rev. *96*, 393-94 (1954).	R. 4-28 MeV N → Emulsion
213	Yu. T. Oganesyan: Range energy dependence of C_{12}, C_{14}, O_{16} in aluminum, copper and gold in the energy interval from 50 to 110 MeV. Zh Eks. Teor. Fiz. *36*, 936-37 (1959) [Engl. trans. Sov. Phys. JETP *9*, 661-62 (1959)].	R. 50-110 MeV ^{12}C, ^{14}C, ^{16}O → Al, Cu, Au
214	H. H. Heckmann, B. L. Perkins, W. G. Simon, F. M. Smith, W. H. Barkas: Ranges and energy-loss processes of heavy ions in emulsion. Phys. Rev. *117*, 544-56 (1960).	R. 0.585 MeV p, 4.18 - 113.4 MeV C, 13.13 - 133.5 MeV N, 5.61 - 153 MeV O, 7.41 - 203.1 MeV Ne, 9.27 - 329.4 MeV Ar → Emulsion
215	J. T. Park: Stopping cross sections of some hydrocarbon gases for 20-400 keV helium ions. Phys. Rev. *138*, A1317-21 (1965).	S. 40-200 keV He → CH_4, C_2H_2, C_2H_4, C_3H_8, C_3H_6, $(CH_2)_3$
216	A. R. Sattler: Ionization produced by energetic silicon atoms within a silicon lattice. Phys. Rev. *138*, A1815-21 (1965).	S. 21.2 - 3139 keV Si → Si

217	R. D. Moorhead: Stopping cross sections of low atomic number materials for He$^+$ 65-180 keV. J. Appl. Phys. *36*, 391-96 (1965).	S. 65 - 180 keV p, He → C, He → Al, Cr
218	C. J. Bakker, E. Segrè: Stopping power and energy loss for ion-pair production for 340 MeV protons. Phys. Rev. *84*, 489-92 (1951).	S. rel. to Al and Cu. 340 MeV p → H$_2$, Li, Be,C, Al, Fe, Cu, Ag, Sn, W, Pb, U
219	D. C. Lorentz, F. J. Zimmerman: Stopping of low-energy, H$^+$ and He$^+$ ions in plastic. Phys. Rev. *113*, 1199-1203 (1959).	S. 40-340 keV H, He → Plastics
220	P. G. Roll, F. E. Steigert: Energy loss of heavy ions in nickel, oxygen and nuclear emulsion. Nuclear Physics *17*, 54-66 (1960).	S. 2-10 MeV/amu He, ^{10}B, ^{11}B, C, N, O, F, Ne → O, Ni, Emulsion
221	W. H. Barkas, S. von Friesen: High-velocity range and energy-loss measurements in Al, Cu, Pb, U and emulsion. Nuovo Cimento Supplemento *19*, 41-62 (1961).	R, S rel. to Cu. 750 MeV p → Al, Cu, Pb, U, Emulsion
222	V. P. Zrelov, G. D. Stoletov: Range-energy relation for 660 MeV protons. Zh. Eksp. Teor. Fiz. *36*, 664-72 (1959) [Engl. trans. Sov. Phys. JETP *9*, 461-67 (1959)].	R. 660 MeV p → Cu. S rel. to Cu, 635 MeV p → H, Be, C, Fe, Cd, W
223	K. Bethge, P. Sandner: Zum Energieverlust schwerer Ionen. Phys. Letters *19*, 241-43 (1965).	S. 5-20 MeV B, 7-28 MeV N → Ag, Ni, Au
224	B. R. Appleton, C. Erginsoy, H. E. Wegner, W. M. Gibson: Axial and planar effects in the energy loss of protons in silicon single crystals. Phys. Letters *19*, 185-86 (1965).	S, δS. 4.85 MeV p → Si (cryst.)
225	C. J. Andreen, R. L. Hines: Channeling of 13 keV O$^+$ ions in gold crystals. Phys. Letters *19*, 116-18 (1965).	S. 13 keV O → Au (cryst.)
226	C. Chasman, K. W. Jones, R. H. Ristinen: Measurement of the energy loss of germanium atoms to electrons in germanium at energies below 100 keV. Phys. Rev. Letters *15*, 245-48 (1965) (Erratum; Phys. Rev. Letters *15*, 684, 1965).	S, $\eta(\varepsilon)$. 20-100 keV Ge → Ge
227	P. M. Portner, R. B. Moore: A precise measurement of the range of 100 MeV protons in aluminum. Can. J. Phys. *43*, 1904-14 (1965).	R. 100 MeV p → Al
228	J. P. S. Pringle: Range profiles for ions implanted into anodic tantalum oxide. J. Electrochem. Soc. *121*, 45-55 (1974).	R. 0.5-160 keV ^{24}Na, ^{42}K, ^{86}Rb, ^{125}Xe, ^{134}Cs, ^{204}Tl, ^{222}Rn → Ta$_2$O$_5$
229	J. W. Boring, G. E. Strohl, F. R. Woods: Total ionization in nitrogen by heavy ions of energies 25 to 50 keV. Phys. Rev. *140*, A1065-69 (1965).	S. 25-50 keV H, He, C, N, O, Ar → N$_2$
230	G. L. Cano, R. W. Dressel: Energy loss and resultant charge of recoil particles from alpha disintegration in surface deposits of ^{210}Po and ^{241}Am. Phys. Rev. *139*, A1883-92 (1965).	R. 103 keV ^{206}Pb, 924 keV ^{239}Np → Ar
231	R. P. Henke, E. V. Benton: Range-momentum relation for heavy recoil ions in emulsion. Phys. Rev. *139*, A2017-21 (1965).	R. 32-320 MeV ^{108}Ag, 30-260 MeV ^{80}Br → Emulsion

232	C. A. Sauter, E. J. Zimmermann: Stopping cross sections of carbon and hydrocarbon solids for low-energy protons and helium ions. Phys. Rev. *140*, A490-98 (1965).	S. 30-350 keV p, He $\rightarrow$ C, plastics
233	J. Williamson, D. E. Watt: The influence of molecular binding on the stopping power of alpha particles in hydrocarbons. Phys. Med. Biol. *17*, 486-92 (1972).	S. 1.5 MeV $\alpha \rightarrow$ C, H, C_2H_6, C_4H_{10}, C_2H_4, C_3H_6, C_4H_8, C_2H_2, C_3H_4, C_4H_6, CH_4, C_6H_6, ethylene, polyethylene, propylene, polypropylene
234	J. A. Panontin, L. L. Schwartz, A. F. Stehney, E. P. Steinberg, L. Winsberg: Ranges of C^{11} in aluminum. Phys. Rev. *140*, A151-55 (1965).	R. 0.65 - 1.64 MeV $^{11}C \rightarrow$ Al
235	P. H. Barker, W. R. Phillips: The range of nitrogen ions in nickel and silver. Proc. Phys. Soc. *86*, 379-85 (1965).	R. 0.4 - 2.5 MeV N $\rightarrow$ Ni, Ag
236	O. Hollricher: Umladnng und Ionisation beim Durchgang von leichten und schweren Wasserstoffionen durch leichten und schweren Wasserstoff. Z. Physik *187*, 41 (1965).	S. 1.5 - 30 keV p, d $\rightarrow$ H_2, D_2
237	H. Lutz, R. Schuckert, R. Sizmann: The ranges of fast heavy particles in solids and theoretical results. Nucl. Instr. Methods *38*, 241-44 (1965).	R, δR. 70 keV $^{85}Kr \rightarrow$ Au (cryst.)
238	H. C. Hayden, R. C. Amme: Low energy ionization of argon atoms by argon atoms. Phys. Rev. *141*, 30-31 (1966).	S. 30 - 2900 eV Ar $\rightarrow$ Ar
239	W. P. Jesse, J. Saduakis: Recoil particles from Po^{210} and their ionization in argon and helium. Phys. Rev. *102*, 389-90 (1956).	R. 103 keV $^{206}Pb \rightarrow$ He, Ar
240	E. M. Zarutskii: Passage of potassium ions through copper and silver films. Fiz. Tverd. Tela *6*, 3734-36 (1964) [Engl. trans. Sov. Phys. Solid State *6*, 2995-96 (1965)].	R. 6-14 keV K $\rightarrow$ Cu, Ag
241	J. L. Whitton: Removal of thin (20Å) layers of metals, metal oxides and ceramics by mechanical polishing. J. Appl. Phys. *36*, 3917-22 (1965).	R, δR. 40 keV $^{133}Xe \rightarrow$ UO_2, ZrO_2, Ta_2O_5, 40 keV $^{134}Cs \rightarrow$ ZrO_2
242	V. Subramanyam, M. Kaplan: Stopping of energetic copper ions in aluminum. Phys. Rev. *142*, 174-78 (1966).	R, δR. 3.35 - 17.6 MeV ^{60}Cu, $^{61}Cu \rightarrow$ Al
243	J. F. Gibbons, A. El-Hoshy, K. E. Manchester, F. L. Vogel: Implantation profiles for 40 keV phosphorous ions in silicon single-crystal substrates. Appl. Phys. Letters *8*, 46-48 (1966).	R, δR. 40 keV $^{32}P \rightarrow$ Si (cryst.)
244	R. B. J. Palmer: The stopping of hydrogen and hydrocarbon vapours for alpha particles over the energy range 1-8 MeV. Proc. Phys. Soc. *87*, 681-88 (1966).	S. 1-8 MeV $\alpha \rightarrow$ H, C, many hydrocarbons
245	H. Bilger, E. Baldinger, W. Czaja: Ionization von Si-Rückstosskernen in Si-Zähldioden bei Bestrahlung mit Neutronen von 3.0 bis 3.9 MeV. Helv. Phys. Acta *36*, 405-12 (1963).	S, $\eta(\varepsilon)$. 430 - 560 keV Si $\rightarrow$ Si

246	A. R. Sattler, F. L. Vook, J. H. Palms: Ionization produced by energetic germanium atoms within a germanium lattice. Phys. Rev. *143*, 588-95 (1966).	S, $\eta(\varepsilon)$. 21-997 keV Ge → Ge
247	B. Fastrup, P. Hvelplund, C. A. Sautter: Stopping cross section in carbon of 0.1-1.0 MeV atoms with $6 \leq Z_1 \leq 20$. Kgl. Danske Videnskab. Selskab. Mat. Fys. Medd. *35*, No. 10, 1-28 (1966).	S. 150 keV H, 82 - 380 keV ^{12}C, 73 - 418 keV ^{14}N, 81 - 479 keV ^{16}O, 138 - 473 keV ^{19}F, 81 - 946 keV ^{20}Ne, 90 - 898 keV ^{23}Na, 135 - 766 keV ^{25}Mg, 88 - 875 keV ^{27}Al, 133 - 780 keV ^{28}Si, 37- 849 keV ^{31}P, 168 - 753 keV ^{32}S, 134-1133 ^{35}Cl, 138 - 1290 keV ^{40}Ar, 138 - 1138 keV ^{39}K, 191 - 874 keV ^{40} Ca → C
248	D. I. Porat, K. Ramavataram: The energy loss of helium and nitrogen ions in metals. Proc. Roy. Soc. *A252*, 394-410 (1959).	S. 0.6 - 0.95 MeV He → Al, 0.4 - 0.95 MeV He → Ni, Ag, 0.35 - 0.95 MeV He → Au, 0.46 - 1.5 MeV N₂ → Al 0.6 - 1.8 MeV N₂ → Ni, Au
249	D.I. Porat, K. Ramavataram: The energy loss and ranges of carbon and oxygen ions in solids. Proc. Phys. Soc. *77*, 97-102 (1961).	S. 0.36 - 3.2 MeV O, C → C, Al, Ni, Ag, Au
250	D. I. Porat, K. Ramavataram: Differential energy loss and ranges of Ne, N, and He ions. Proc. Phys. Soc. *78*, 1135-43 (1961).	S. 0.4 - 6.2 MeV Ne → C, Al, Ni, Ag, Au, 0.36 - 2.6 MeV N₂ → C, 0.4 - 5.2 MeV N₂ → Al, 0.4 - 5.6 MeV N₂ → Ni, 0.4 - 3.8 MeV N₂ → Ag, Au, 0.3 - 1.3 MeV He → C, 0.3 - 2.0 MeV He → Al, Ni, Ag, Au, 0.4 - 1.0 MeV D₂ → Ag
251	E. Rotondi, K. W. Geiger: The measurement of range-energy relation and straggling of α-particles in air using a solid state detector. Nucl. Instr. Meth. *40*, 192-96 (1966).	R, δR. 0.5 - 5.6 MeV α → Air
252	A. R. Sattler, G. Dearnaley: Anomalous energy losses of protons channeled in single crystal germanium. Phys. Rev. Letters *15*, 59-61 (1965).	S, δS. 4.25 - 7.75 MeV p, 7.63 MeV d → Ge (cryst.)
253	D. M. Parfanovich, A. M. Semchinova, G. N. Flerov: Determination of the range-energy relation for nitrogen and oxygen ions in photographic emulsion. Zh. Eksp. Teor. Fiz. *33*, 343-45 (1957) [Engl. trans. Sov. Phys. JETP *6*, 266-67 (1958)].	R. 3-120 MeV N, O → Emulsion
254	J. P. Hazan, M. Blann: Excitation functions, recoil ranges and statistical theory of analysis of reactions induced in Fe^{56} with 6-29-MeV He^3 ions. Phys. Rev. *137*, B1202-13 (1965).	R. 0.5-4.5 MeV Co, Ni → Au
255	M. Kaplan, R. D. Fink: Recoil properties of Sm^{142} from nuclear reactions induced by heavy ions. Phys. Rev. *134*, B30-32 (1964).	R. 2-12 MeV ^{142}Sm → Al

256	M. Kaplan, J. L. Richards: Ranges of Ba^{126} and Ba^{128} recoil fragments in aluminum. Phys. Rev. *145*, 153-57 (1966).	R, δR. 2.8 - 14.2 MeV ^{126}Ba, ^{128}Ba → Al
257	B. R. Appleton, M. Altman, L. C. Feldman, W. M. Gibson, C. Erginsoy: Least energy loss and its dispersion for "channeled" protons in silicon and germanium single crystals. Bull. Am. Phys. Soc. *11*, 176 (1966).	S, δS. 3-11 MeV p → Si, Ge (both cryst.)
258	H. O. Lutz, S. Datz, C. D. Moak, T. S. Noggle, L. C. Northcliffe: Charge-state distribution of channeled 40 MeV ^{127}I ions in Au single crystals. Bull. Am. Phys. Soc. *11*, 177 (1966).	S. 40 MeV I → Au (cryst.)
259	P. M. Mulas, R. C. Axtmann: Energy loss of fission fragments in light materials. Phys. Rev. *146*, 296-300 (1966).	S. Fiss. fragm. → H$_2$, D$_2$, Mylar, N$_2$
260	V. A. J. van Lint, M. E. Wyatt, R. A. Schmitt, C. S. Suffredini, D. K. Nichols: Range of photoparticle recoil atoms on solids. Phys. Rev. *147*, 242-48 (1966).	R. 7 x 10^{-4} < ε < 5 Ti, Sc, Cr, Fe, Mn, Ni, Co, Ge, Zr, Y, Sr, Mo, Rh, Pd, Ag, Cd, Sn, Gd, Ta, Au, Th → Al, Cu (specific combinations not identifiable)
261	D. Marx: Messung des Bremsvermögens von Pb^{208} - Alpha - Rückstoss-Kernen in fester Materie. Z. Physik *195*, 26-43 (1966).	S. 169 keV ^{208}Pb → C, Ag, Au
262	O. Selig, R. Sizmann: Die Reichweiteverteilung von Spaltprodukten in Feskörpern. Nukleonik *8*, 303-14 (1966).	R, δR. 97 MeV ^{95}Zr, 65 MeV ^{140}Ba → Mica, Al (cryst.)
263	J. A. Davies, P. Jespersgard: Anomalous penetration of xenon in tungsten crystals - A diffusion effect. Can. J. Phys. *44*, 1631-38 (1966).	R, δR. 20 keV Xe → W (cryst.)
264	K. Bethge, P. Sandner, H. Schmidt: Energieverluste und Ladungszustände schwerer Ionen beim Durchgang durch Materie. Z. Naturforschg. *21a*, 1052-57 (1966).	S. 5-20 MeV B, 5-30 MeV O, 7-28 MeV N, 5-30 MeV S → Ni, Ag, Au
265	H. Hermann, H. Lutz, R. Sizmann: Zur Eindringtiefe von 70 keV - Krypton - Ionen in Wolfram-Einkristallen. Z. Naturforschg. *21a*, 365-66 (1966).	R, δR. 70 keV ^{85}Kr → W (cryst.)
266	J. R. MacDonald, J. H. Ormrod, H. E. Duckworth: Stopping cross section in boron of low atomic number atoms with energies from 15 to 140 keV. Z. Naturforschg. *21a*, 130-34 (1966).	S. 12-65 keV, H, D, 15-70 keV He, Li, 15-140 keV B, C, N. O, F, 20-140 keV Ne, 25 - 70 keV Na → B
267	H. O. Lutz, S. Datz, C. D. Moak, T. S. Noggle: Determination of interatomic potentials and stopping powers from channeled-ion energy-loss spectra. Phys. Rev. Letters *17*, 285-87 (1966).	S, δS. 60 MeV I, 3 MeV α → Au (cryst.)
268	K. Hosono, R. Ishiwari, Y. Uemura: Measurement of absolute energy loss of 28 MeV alpha particles in various materials. Bull. Inst. Chem. Res. Kyoto Univ. *43*, 323-29 (1965).	S. 28 MeV α → Au, Sn, mylar

269	H. H. Andersen, A. F. Garfinkel, C. C. Hanke, H. Sørensen: Stopping power of aluminum for 5-12 MeV protons and deuterons. Kgl. Danske Videnskab. Selskab Mat. Fys. Medd. *34*, No. 4, 1-24 (1966).	S. 5-12 MeV p, d → Al
270	C. D. Moak, M. D. Brown: Some heavy-ion stopping powers. Phys. Rev. *149*, 244-45 (1966).	S. 10-100 MeV Br, I → Be, C, Al, Ni, Ag, Au
271	R. D. Schuckert, H. Lutz, R. Sizmann: Angular dependence of channeling in gold crystals. Z. Naturforschg. *21a*, 1296-98 (1966).	R, δR. 70 keV ^{85}Kr → Au (cryst.)
272	L. Erikson, J. A. Davies, P. Jespersgaard: Range measurements in oriented tungsten single crystals (0.1-1.0 MeV). Part I: Electronic and nuclear stopping powers. Phys. Rev. *161*, 219-34 (1967).	R, δR. 0.1-1.0 MeV ^{24}Na, ^{32}P, ^{42}K, ^{51}Cr, ^{64}Cu, ^{82}Br, ^{85}Kr, ^{86}Rb, ^{122}Sb, ^{125}Xe, ^{133}Xe, ^{187}W, ^{222}Rn → W (cryst.) 40-500 keV ^{24}Na, 150-500 keV ^{42}K, 20-160 keV ^{85}Kr 40-500 keV ^{133}Xe → Al (cryst.)
273	J. Kloppenburg, A. Flammersfeld: Energieverlustmessungen in Antrazen, Terphenyl und Plastikzintillatoren für Protonen und Deuteronen in Energiebereich von 100 bis 900 keV. Z. Physik *196*, 424-32 (1966).	S. 0.1-0.9 MeV p, d → anthrazene, terhenylen, plast. scintillators.
274	J. R. Comfort, J. F. Decker, E. T. Lynk, M. O. Scully, A. R. Quinton: Energy loss and straggling of alpha particles in metal foils. Phys. Rev. *150*, 249-56 (1966).	S, δS. 2-9 MeV α → Al, Ni, Ag, Au
275	F. H. Eisen: Channeling of 375 keV protons through silicon. Phys. Letters *23*, 401-02 (1966).	S, δS. 375 keV p → Si (cryst.)
276	J. L. Whitton, Hj. Matzke: The effect of crystallinity and bombardment dose on the penetration of 40 keV xenon ions in ionic crystals and ceramics. Can. J. Phys. *44*, 2905-14 (1966).	R, δR. 40 keV Xe → NaCl, KBr, MgO, SiO_2, UO_2
277	P. J. Walsh, N. Underwood: Stopping powers for Am^{241} alpha particles in gases. Bull. Am. Phys. Soc. *4*, 534 (1966).	S. 0.2-5.5 MeV α → propane, ethylen, acetylene
278	J. C. Overley, W. Whaling: Highly excited states in C^{11} elastic scattering of protons by B^{10}. Phys. Rev. *128*, 315-24 (1962).	S. 0.1-3.0 MeV p → B
279	S. Gorodetzky, A. Chevallier, A. Pape, J. C. Sers, A. M. Bergdolt, M. Burs, R. Armbruster: Mesure des pouvoirs d'arrêt de C, Ca, Au et CaF$_2$ pours des protons d'énergie comprise entre 0.4 et 6 MeV. Nuclear Physics *A91*, 133-44 (1967).	S. 0.4-6.0 MeV p → C, Ca, Au, CaF_2
280	H. H. Andersen, C. C. Hanke, H. Sørensen and P. Vajda: Stopping power of Be, Al, Cu, Ag, Pt and Au for 5-12 MeV protons and deuterons. Phys. Rev. *153*, 338-42 (1967).	S. 4.5 - 12 MeV p, d → Be, Al, Cu, Ag, Pt, Au
281	D. Kamke, P. Kramer: Energieverlust und Reichweite von α-Teilchen in Bor im Energiebereich von 0.2 bis 5.3 MeV. Z. Physik *168*, 465-73 (1962).	S. 0.2 - 5.3 MeV α → B

282	D. L. Mason, R. M. Prior, A. R. Quinton: The energy straggling of 1 MeV protons in gases. Nucl. Instr. Meth. *45*, 41-44 (1966).	δS. 1 MeV p → H, He, N, O, Ar, Xe
283	C. Chasman, K. W. Jones, R. A. Ristinen, J. T. Sample: Measurement of energy loss of germanium atoms to electrons in germanium at energies below 100 keV. II. Phys. Rev. *154*, 239-44 (1967).	S, $\eta(\varepsilon)$. 20-100 keV Ge → Ge
284	C. J. Andreen, R. L. Hines: Channeling of D^+ and He^+ ions in gold crystals. Phys. Rev. *151*, 341-48 (1966).	S. 15 keV d, α → Au and Au (cryst.)
285	L. B. Bridwell, C. D. Moak: Stopping power and differential ranges for ^{79}Br and ^{127}I in UF_4. Phys. Rev. *156*, 242-43 (1967).	S. 20-100 MeV ^{79}Br, ^{127}I → UF_4
286	J. A. Davies, L. Erikson, P. Jespersgaard: The range of heavy ions (0.1 - 1.5 MeV) in monocrystalline tungsten. Nucl. Instr. Methods *38*, 245-48 (1965).	R, δR. 0.1 - 1.5 MeV Na, P, K, Kr, Xe → W (cryst.)
287	J. L. Whitton: Channelling in gold. Can. J. Phys. *45*, 1947-57 (1967).	R, δR. 20-80 keV ^{133}Xe, ^{198}Au → Au (cryst.)
288	V. V. Makarov, N. N. Petrov: Penetration of light atomic and molecular ions into SiC single crystals. Fiz. Tverd. Tela *8*, 3723-25 (1966) [Engl. trans. Sov. Phys. Solid State *8*, 2993-84 (1967)].	R. 4-20 keV H, H_2, H_3, D, D_2, D_3, He → SiC
289	L. B. Bridwell, L. C. Northcliffe, S. Datz, C. D. Moak, H. O. Lutz: Stopping powers for iodine ions at energies up to 200 MeV. Phys. Rev. *159*, 276-77 (1967).	S. 90-200 MeV I → Be, C, Al, Ni, Ag, Au, UF_4
290	C. J. Andreen, R. L. Hines: Critical angles for channelling of 1 to 25 keV H^+, D^+ and He^+ in gold crystals. Phys. Rev. *159*, 285-90 (1967).	S. 14-28 keV p, d, α → Au, Au (cryst.)
291	K. Morita, H. Akimura, T. Suita: Stopping cross-sections of metallic films for projectile of low energy proton. J. Phys. Soc. Jap. *22*, 1503 (1967).	S. 7-35 keV p → Be, Al, Cu, Ag, Au
292	C. A. Nicoletta, P. L. McNulty, P. L. Jain: Ionization loss as a function of energy by four different proton beams in the same emulsion. Bull. Am. Phys. Soc. *12*, 28 (1967).	S. 5-24 GeV p → Emulsion
293	L. B. Bridwell, L. C. Northcliffe, S. Datz, C. D. Moak, H. O. Lutz: Stopping power of C, Al, Ni, Ag, Au, and UF_4 for 10-200 MeV ^{127}I ions. Bull. Am. Phys. Soc. *12*, 28 (1967).	S. 10-200 MeV ^{127}I → C, Al, Ni, Ag, Au, UF_4
294	P. Hvelplund: Specialeopgave. Aarhus University (1967) (see ref. 315). (In Danish)	S. 200-1200 keV heavy ions → C
295	L. Hastings, P. R. Ryall, A. van Wijngaarden: The energy loss of heavy ions in ZnS:Ag in the keV range. Can. J. Phys. *45*, 2334-42 (1967).	S. 5-80 keV H, 10-100 keV He, 12-80 keV N, 15-90 keV Ar, 25-90 keV Kr → ZnS:Ag
296	W. Wiechmann, J. Biersack: Reichweiten von Radium - 224 - Rückstoss-atomen in Gasen. Nukleonik *9*, 399-400 (1967).	R, δR. 96 keV ^{224}Ra → Ar, Ne, O_2, N_2, CH_4, He, H_2

297	G. S. Anderson: Etching rate of an ion-barded tungsten (110) surface. J. Appl. Phys. *38*, 1989-91 (1967).	R. 5 keV Kr → W
298	B. Domeij, I. Bergström, J. A. Davies, J. Uhler: A method of determining heavy ion ranges by analysis of α-line shapes. Arkiv för Fysik *24*, 399-411 (1963).	R, δR. ^{222}Rn 140-210 keV → Al, 70-210 keV → W, 133 keV → Ag, Au
299	C. K. Valentine, M: Blann: Mean ranges of 0.2 to 5 MeV nickel and cobalt isotopes in iron. Bull. Am. Phys. Soc. *12*, 29 (1967).	R. 0.2-5.0 MeV Ni, Co → Fe
300	A. Sperduto, W. W. Buechner, R. J. van de Graaff: Range-energy measurements for heavy ions. Bull. Am. Phys. Soc. *12*, 28 (1967).	R. 100-240 MeV F, Br, I, Ta, U → Emulsion
301	R. W. Bower, R. Baron, J. W. Mayer, O. J. Marsh: Deep (1-10μ) penetration of ion-implanted donors in silicon. Appl. Phys. Letters *9*, 203-05 (1966).	R, δR. 20 keV Sb → Si (cryst.)
302	L. Erikson: Range measurements in oriented tungsten single crystals (0.1-1.0 MeV). Part II: A detailed study of the channeling of K^{42} ions. Phys. Rev. *161*, 235-44 (1967).	R, δR. 0.07 - 1.0 MeV ^{42}K → W (cryst.)
303	D. A. Channing: Effect of temperature on the penetration of heavy keV ions in monocrystalline solids. I. ^{133}Xe ions in gold. Can. J. Phys. *45*, 2455-66 (1967).	R, δR. ^{133}Xe 40 keV → Au (cryst.)
304	L. M. Howe, D. A. Channing: Effect of temperature on the penetration of heavy keV ions in monocrystalline solids. II. Various ions in Au, Al and W. Can. J. Phys *45*, 2467-82 (1967).	R, δR. 40-94 keV ^{198}Au, 40 keV ^{85}Kr, ^{24}Na → Au (cryst.), 40 keV ^{133}Xe, 40-65 keV ^{24}Na → Al (cryst.), 40 keV ^{133}Xe, ^{24}Na → W (cryst.)
305	B. R. Appleton, C. Erginsoy, W. M. Gibson: Channeling in the energy loss of 3-11 MeV protons in silicon and germanium single crystals. Phys. Rev. *161*, 330-49 (1967).	S. 3-11 MeV p → Si, Ge (both cryst.). Chann. and random
306	J. B. Cumming, V. D. Crespo: Energy loss and range of fission fragments in solid media. Phys. Rev. *161*, 287-93 (1967).	S. Fiss. fragm. → Mylar
307	A. R. Sattler: Velocity and charge dependence of the energy losses of the channeling peak. Bull. Am. Phs. Soc. *12*, 392 (1967).	S. 3-4 MeV p, d, α → GaSb (cryst.)
308	A. R. Sattler, G. Dearnaley: Channeling in diamond-type and zinc-blende lattices: Comparative effects in channeling of protons and deuterons in Ge, GaAs, and Si. Phys. Rev. *161*, 244-52 (1967) (Erratum; Phys. Rev. *165*, 750 (1968)).	S. 4-7.6 MeV p, d → Ge, GaAs, Si (all cryst.)

309	W. W. Bowman, F. M. Lanzafame, C. K. Cline, Yu-Wen Yu, M. Blann: Recoil ranges of 0.22 - 5.2 MeV ions in vanadium, nickel, iron, zirconium and gold. Phys. Rev. *165*, 485-93 (1968).	R. ~1-5 MeV ^{54}Mn, ^{55}Mn, ^{51}Cr, ^{48}V, ^{47}Sc, ^{46}Sc → ^{51}V, ~0.2 - 0.7 MeV ^{55}Co, ^{56}Co, ^{57}Co, ^{58}Co, ^{57}Ni → Natl. Ni, ~2-5 MeV ^{56}Co, ^{57}Co, ^{57}Ni → ^{54}Fe, ~2 - 3.5 MeV ^{90}Mo, ^{98}Nb, ^{89}Zr, ^{88}Zr, ^{87}Zr, ^{86}Zr, ^{86}Y, ^{88}Y → ^{90}Zr, Natl. Zr, ~0.4 - 1.5 MeV ^{196}Tl, ^{197}Tl, ^{198}Tl, ^{199}Tl, ^{200}Tl → ^{197}Au
310	D. Powers, W. K. Chu, P. D. Bourland: Range of Ar, Kr, and Xe ions in solids in the 500 keV to 2 MeV energy region. Phys. Rev. *165*, 376-87 (1968).	R, δR. 0.5 - 2.0 MeV Ar, Kr, Xe → Be, C; Kr, Xe → Al; Xe → V, Ni, Cu; S. 0.6 - 2.0 MeV p → V
311	N. O. Lassen, N. O. Roy Poulsen, G. Sidenius, L. Vistisen: Stopping of 50 keV ions in gases. Kgl. Danske Videnskab. Selskab Mat. Fys. Medd. *34*, No. 5, 1-20 (1964).	R. 50 keV ^{24}Na, ^{66}Ga, ^{198}Au → H_2, D_2, Ne, Ar; 50 keV ^{66}Ga → ^{3}He, ^{4}He, N_2
312	A. Marcinkowski, H. Rzewuski, Z. Werner: Range-energy relation for low energy protons in Si and Ge. Nucl. Instr. Meth. *57*, 338-40 (1967).	R. 0.8 - 1.9 MeV p → Ge, Si
313	I. M. Vasilievsky, Yu. D. Prokoshkin: Ionization energy loss of protons, denterons and α-particles. Yaderna Fiz. (SSSR) *4*, 549-55 (1966) [Engl. trans. Sov. Phys. Nucl. Phys. *4*, 390-94 (1967)].	S. 267, 650 MeV p, 377 MeV d, 765 MeV α → Cu, S rel. to Cu same proj. → H, C, Al, Sn, Pb
314	H. D. Maccabee, M. R. Raju, C. H. Tobias: Fluctuations of energy loss by heavy charged particles in thin absorbers. Phys. Rev. *165*, 469-74 (1968).	δS. 45, 730 MeV p, 910 MeV α, 370 MeV π^- → Si
315	P. Hvelplund, B. Fastrup: Stopping cross section in carbon of 0.2 - 1.5 MeV atoms with $21 \leq Z_1 \leq 39$. Phys. Rev. *165*, 408-14 (1968).	S. 230 - 1170 keV Sc, 240 - 970 keV Ti, 390 - 1485 keV Cr, 300 - 1170 keV Mn, 238 - 1470 keV Fe, 240 - 1185 keV Co, 400 - 1485 keV Cu, 395 - 1485 keV Ge, 590 - 1480 keV Br, 590 - 1485 keV W, 490 - 1480 keV Y → C
316	A. H. Morton, D. A. Aldcroft, M. F. Payne: Energy loss by low-energy protons in gold. Phys. Rev. *165*, 415-19 (1968).	S. 10-50 keV p → Au
317	G. F. Bogdanov, V. P. Kabaev, F. V. Lebedev, G. M. Noviko: Stopping power of nickel for protons and He ions for the energy range 20-95 keV. Atomnaya Energiya (SSSR) *22*, 126-27 (1967) [Engl. trans. Soviet Atom. Energy *22*, 133-34, (1967)].	S. 20-95 keV p, α → Ni
318	E. M. Zarutskii: Penetration of hydrogen ions into copper. Fiz. Tverd. Tela *9*, 1500-04 (1967). [Engl. trans. Sov. Phys. Solid State *9*, 1172-76 (1967)].	S. 4-20 keV H$^+$ → Cu
319	S. Kahn, V. Forgue: Range-energy relation and energy loss of fission fragments in solids. Phys. Rev. *163*, 290-96 (1967).	S. Fiss. fragm. → Al, Ni, Ag, Au, U

320	M. Mannami, F. Fujimoto, K. Ozawa: Anomalous energy losses of 1.5 MeV protons channeled in silicon single crystals. Phys. Letters *26A*, 201-02 (1968).	S, δS. 1.5 MeV p $\rightarrow$ Si (cryst.)
321	E. Bellamy, R. Hofstadter, W. L. Lakin, J. Cox, M. I. Perl, W. T. Tower, Z. I, Zipf: Energy loss and straggling of high-energy muons in NaI. Phys. Rev. *164*, 417-20 (1967).	S, δS. 0.5-10.5 GeV/c μ^+, μ^- $\rightarrow$ NaI
322	B. H. Armitage, B. W. Hooton: Energy loss of oxygen and sulphur ions in matter. Nucl. Instr. Meth. *58*, 29-35 (1968).	S. 10-30 MeV O, 19-40 MeV S $\rightarrow$ Ag, Au
323	B. de Cosnac, P. Dulien, J. P. Noël: Mesure de L'Énergie Cédée au Réseau par un Primaire dans le Silicium. Rev. Physique Appl. *2*, 158-63 (1967).	S, $\eta(\varepsilon)$. 100-400 keV Si $\rightarrow$ Si
324	W. White, R. M. Mueller: Measurement of atomic-stopping cross sections at low energies. J. Appl. Phys. *38*, 3660-61 (1967).	S. 20-140 keV p, α $\rightarrow$ Al
325	L. Hastings, A. van Wijngarden: The energy loss, the detorication depth and the light output for heavy ions in ZnO:Zn. Can. J. Phys. *45*, 4039-51 (1967).	S rel. to p. 10-100 keV α, N, Ar, Kr $\rightarrow$ ZnO:Zn
326	J. M. Bewers, F. C. Flack: Stopping power and the additivity rule for some fluorine compounds. Nucl. Instr. Methods *59*, 337-38 (1968).	S. 1 MeV p $\rightarrow$ CaF_2, LiF, Na_2SiF_6, KBF_4, NaF, K_2SiF_6, $BaSiF_6 \cdot H_2O$, $(NH_4)_2SiF_6$, KF, PbF_2, $ZnF_2 \cdot 3H_2O$, CdF_2, $ZnSiF_6 \cdot 4H_2O$, $NiF_2 \cdot 15H_2O$, $CuF_2 \cdot 2H_2$, BaF_2
327	L. Lehmann, H. Spehl, N. Wertz: A new method for range measurements of very heavy ions using Coulomb excitation and perturbed angular correlations. Nucl. Instr. Methods *55*, 201-04 (1967).	R. 0.8 MeV W $\rightarrow$ Cu
328	J. L. Whitton: The measurement of ionic mobilities in the anodic oxides of tantalum and zirconium by a precision sectioning technique. J. Electrochem. Soc. *115*, 58-61 (1968).	R, δR. 30 keV ^{82}Br, ^{85}Kr, ^{86}Rb $\rightarrow$ ZrO_2, Ta_2O_5, 30 keV ^{125}Xe, ^{133}Xe $\rightarrow$ ZrO_2
329	T. M. Duc, A. Demeyer, J. Tousset, R. Chery: Détermination Expérimentale de la Perte d'Énergie, des Parcours et de la Dispersion d'un Faisceau de Particules Alpha de 54.5 MeV dans Quelques Éléments. J. Physique *29*, 129-135 (1968).	R, S, δS. 54.4 MeV α $\rightarrow$ Cu, Ag, Tb, Tm, Au, S, δS.50-54 MeV α, 27 MeV d $\rightarrow$ Al
330	P. D. Croft, K. Street, Jr: Range-energy studies of Po and At recoils in Al and Al_2O_3. Phys. Rev. *165*, 1375-80 (1968).	R. 6-12 MeV Po, At $\rightarrow$ Al, Al_2O_3.
331	G. L. Cano: Total ionization and range of low-energy recoil particles in pure and binary gases. Phys. Rev. *169*, 277-79 (1968).	R. 103 keV ^{206}Pb $\rightarrow$ Ne, Ar, Xe, N_2, Air, Hydrocarbons

332	J. P. Biersack: Range of recoil atoms in isotropic stopping materials. Z. Physik *211*, 495-501 (1968).	R. 96 keV ^{224}Ra → Ar, Ne, O_2, N_2, CH_4, He, H_2; 1335 keV ^{24}Na → Al; 328 keV ^{27}Mg → Al; 219 keV ^{56}Mn → Fe; 145 keV ^{58}Co → Ni, 0.5 keV ^{64}Cu → CuO; 0.6 keV ^{28}Al → Al_2O_3
333	W. J. Kleinfelder, W. S. Johnson, J. F. Gibbons: Impurity distribution profiles in ion-implanted silicon. Can. J. Phys. *46*, 597-606 (1968).	R, δR. 10-70 keV B, N, P, As → Si (cryst.)
334	G. Dearnaley, J. H. Freeman, G. A. Gard, M. A. Williams: Implantation profiles of ^{32}P channeled into silicon crystals. Can. J. Phys. *46*, 587-95 (1968).	R, δR. 10-110 keV ^{32}P → Si (cryst.)
335	J. L. Whitton: The depth distribution of 40 keV ^{133}Xe ions in various single crystals. Can. J. Phys. *46*, 581-86 (1968).	R, δR. 40 keV ^{133}Xe → Ta, W, Al, Cu, Au, Ir (all cryst.)
336	J. A. Davies, L. Erikson, J. L. Whitton: Range measurements in oriented tungsten single crystals. III. The influence of temperature on the maximum range. Can. J. Phys. *46*, 573-79 (1968).	R, δR, R_{max}. 40 keV ^{42}K → W (cryst.)
337	N. J. Freeman, I. D. Latimer: The ranges of 5-80 keV deuterium ions in gold and aluminum. Can. J. Phys. *46*, 467-72 (1968).	R. 5-80 keV d → Al, Au
338	T. Andersen, G. Sørensen: Range studies using a new chemical film technique. Can. J. Phys. *46*, 483-88 (1968).	R, δR. 100-550 keV ^{24}Na, 150-500 keV ^{32}P, 100-500 keV ^{42}K → Au
339	F. H. Eisen: Channeling of medium-mass ions through silicon. Can. J. Phys. *46*, 561-72 (1968).	S. 100-500 keV B, C, N, O, F, Ne, Na, Mg, Al, Si, P, Cl, Ar, K → Si (cryst.)
340	R. Eldros, R. L. Hines: Energy losses of channeled D^+ ions in gold crystal foils from 16 to 30 keV. Bull Am. Phys. Soc. *13*, 402 (1968).	S. 16-30 keV d → Au (cryst.)
341	B. Fastrup, A. Borup, P. Hvelplund: Stopping cross section in atmospheric air of 0.2-0.5 MeV atoms with $6 \leq Z_1 \leq 24$. Can. J. Phys. *46*, 489-95 (1968).	S. 100-500 keV ^{12}C, ^{14}N, ^{16}O, ^{20}Ne, ^{23}Na; 200-500 keV ^{24}Mg, ^{31}P, ^{32}S, ^{35}Cl, ^{45}Sc, 250-500 keV ^{40}Ca, ^{48}Ti; 200-400 keV ^{27}Al; 150-1000 keV ^{40}Ar, 200-800 keV ^{39}K, 300-800 keV ^{52}Cr → Air
342	J. H. Ormrod: Low-energy electronic stopping cross sections in nitrogen and argon. Can. J. Phys. *46*, 497-502 (1968).	S. 5-100 keV H, 10-50 keV D, 10-90 keV He, 20-160 keV B, 20-180 keV C, 20-200 keV N, O, 30-140 keV F, 20-180 keV Ne → N; 5-80 keV H, 10-80 keV D, He, 25-160 keV B, C, N, O, F, Ne → Ar

343	W. M. Gibson, J. B. Rasmussen, P. A. Olesen, C. J. Andréen: Charged-particle energy loss in thin gold crystals. Can. J. Phys. *46*, 551-60 (1968) [Erratum; Can. J. Phys. *47*, 1756 (1969)].	S, δS. 400 keV H, 800 keV He $\rightarrow$ Au (cryst.)
344	D. H. Poole, A. G. Warner, R. Hancock, R. L. Woodley: Stopping cross-section measurements in carbon using recoil from radioactive decay. Brit. J. Appl. Phys. *1*, 309-12 (1968).	S, δS. 169 keV Pb $\rightarrow$ C
345	W. H. Barkas, J. N. Dyer, H. H. Heckman: Resolution of the Σ^--mass anomaly. Phys. Rev. Letters *11*, 26-28 (1963).	R. 12 MeV Σ^+, Σ^- $\rightarrow$ Emulsion
346	E. M. Zarutskii: Energy spectrum of alkali metal ions transmitted by thin copper films. Fiz. Tverd. Tela *9*, 1896-98 (1968) [Engl. trans. Sov. Phys. Solid State *9*, 1495-97 (1968)].	S. 3-20 keV Li, Na, K $\rightarrow$ Cu
347	T. E. Pierce, W. W. Bowman, M. Blann: Stopping power of S^{32}, Cl^{35}, Br^{79} and I^{127} ions in mylar. Phys. Rev. *172*, 287-91 (1968).	S. 15-95 MeV ^{32}S, ^{35}Cl, 30-90 MeV ^{79}Br, 60-105 MeV ^{127}I $\rightarrow$ Mylar
348	K. Bulthuis: Anomalous penetration of Ga and In implanted in silicon. Phys. Letters *27A*, 193-94 (1968).	R, δR. 56 keV In, Ga $\rightarrow$ Si (cryst.)
349	V. G. Volod'ko, E. I. Zorin, P. V. Pavlov, D. I. Tetel'baum: Distribution and range of boron and phosphorus ions used in bombardment of SiO_2. Fiz. Tverd. Tela *10*, 1048-52 (1968) [Engl. trans. Sov. Phys. Solid State *10*, 828-31 (1968)].	R, δR. 30-100 keV B, 50-150 keV P $\rightarrow$ SiO_2
350	W. K. Chu, P. D. Bourland, K. H. Wang, D. Powers: Range and dE/dx of C, N, O, F, and Ne in Be and C from 500 keV to 2 MeV. Phys. Rev. *175*, 342-53 (1968).	R. 0.4-1.9 MeV C, 0.5-2.0 MeV N, 0.3-2.0 MeV O, 0.5-2.0 MeV F $\rightarrow$ Be; 0.2-1.5 MeV O $\rightarrow$ C, 0.5-2.0 MeV Ne $\rightarrow$ Be, C
351	K. Ibel, R. Sizmann: Energy loss of <110> channeled α-recoil atoms in gold. Phys. Stat. Sol. *29*, 403-15 (1968).	S. 169 keV ^{208}Pb, 146 keV ^{210}Pb, 116 keV ^{208}Tl $\rightarrow$ Au (cryst.)
352	I. S. Jones, K. M. Pathak, M. G. Thompson: Ionization energy loss of muons in a plastic scintillator. J. Phys. A: General Phys. *1*, 584-87 (1968).	S. 0.3-120 GeV/c μ^+ Plast. scint.
353	R. Ishiwari, N. Shiomi, Y. Mori, T. Ohata, Y. Uemura: Comparison of energy losses of protons and deuterons of exactly the same velocity. Bull. Inst. Chem. Res. Kyoto Univ. *45*, 379-87 (1967).	S. 7 MeV p, 14 MeV d $\rightarrow$ Al
354	T. E. Pierce, M. Blann: Stopping powers and ranges of 5-90 MeV S^{32}, Cl^{35}, Br^{79}, and I^{127} ions in H_2, He, N_2, Ar, and Kr: A semiempirical stopping power theory for heavy ions in gases and solids. Phys. Rev. *173*, 390-405 (1968).	S. 5-90 MeV ^{32}S, ^{35}Cl, ^{79}Br, ^{127}I $\rightarrow$ H_2, He, N_2, Ar, Kr
355	C. H. Johnson, R. L. Kernell: Use of the (p,n) reaction to measure proton atomic stopping powers in Ag, Cd, In, and Sn. Phys. Rev. *169*, 974-77 (1968).	S. 4.5 MeV p $\rightarrow$ Ag, Cd, In, Sn

356	A. van Wijngaarden, H. E. Duckworth: Energy loss in condensed matter of ^1H, and ^4He in the energy range $4 < E < 30$ keV. Can. J. Phys. *40*, 1749-64 (1962).	S. 4-30 keV H, He $\rightarrow$ C, Al_2O_3
357	P. Jespersgaard, J. A. Davies: Ranges of Na, K, W, and Xe ions in amorphous Al_2O_3 in the energy region 40-1000 keV. Can. J. Phys. *45*, 2983-94 (1967).	R, δR. 40-1000 keV Na, K, Kr, Xe $\rightarrow$ Al_2O_3
358	H. H. Andersen, C. C. Hanke, H. Simonsen, H. Sørensen, P. Vajda: Stopping power of the elements $Z = 20$ through $Z = 30$ for 5 - 12 MeV protons and deuterons. Phys. Rev. *175*, 389-95 (1968).	S. 5-12 MeV p, d $\rightarrow$ Ca, Sc, Ti, V, Cr, Mn, Fe, Co, Ni, Cu, Zn
359	V. N. Lepeshinskaya, E. M. Zarutskii: Penetration of some alkali metal ions into copper and silver. Izv. AN SSSR Ser. fiz. *28*, 1390-95 (1964) [Engl. trans. Bull. Acad. Sci. USSR. Phys. Ser. *28*, 1296-1300 (1964)].	R. 2-12 keV Li, 6-14 keV Na $\rightarrow$ Cu, Ag
360	D.Alexander, K. M. Pathak, M. G. Thompson: Cerenkov energy loss of muons in water. J. Phys. A*1*, 578-83 (1968).	S. 0.3-120 GeV/c μ^+ $\rightarrow$ Plast. scint.
361	E. M. Zarutskii, V. E. Rink: Penetration of lithium and sodium ions into gold. (In Russian). Trudy Leningrad Polytekh. Inst. No. 277, 116-20 (1967).	R. 5-15 keV Li, Na $\rightarrow$ Au
362	Ya. A. Teplova, V. S. Nikolaev, I. S. Dimitriev, L. N. Fateeva: Slowing down of multicharged ions in solids and gases. Zh. Eksp. Teor. Fiz. *42*, 44-60 (1962) [Engl. trans. Sov. Phys. JETP*15*, 31-41 (1962)].	S, R. 75 keV/amu - 1500 keV/amu He, Li, Be, B, C, N, O, Ne, Na, Mg, Al, P, Cl, K, Br, Kr $\rightarrow$ H_2, He, CH_4, benzene, air, Ar, S. Same $\rightarrow$ Al, Ni, Ag, Au
363	K. E. Zimen, D. Ertel: Kernrückstoss in Festkörpern 2. Die Reaktion $Al^{27}(n,p)Mg^{27}$. Nukleonik *4*, 231-32 (1962).	R. 328 keV ^{27}Mg $\rightarrow$ Al
364	D. Ertel, K. Zimen: Kernrückstoss in Festkörpern 3. Die Reaktion $Al^{27}(n,\alpha)Na^{24}$. Nukleonik *5*, 256-58 (1963).	R. 1335 keV ^{24}Na $\rightarrow$ Al
365	R. Oberhauser, W. Wiechmann: Kernrückstoss in Festkörpern 6. Die Reaktion $Ni^{58}(n,p)Co^{58}$, Nukleonik *8*, 59 (1966).	R. 145 keV ^{58}Co $\rightarrow$ Ni
366	H. Nakata: Ranges of nitrogen ions in Al, Ni, Ag, and Au. Can. J. Phys. *46*, 2765-69 (1968) (Erratum; Can. J. Phys. *48*, 1744 (1970)).	R. 1-12 MeV ^{14}N $\rightarrow$ Al, Ni, Ag, Au
367	W. Wiechmann, D. Ertel, K. Zimen: Kernrückstoss in Festkörpern 5. Die Reaktion $Fe^{56}(n,p)Mn^{56}$. Nukleonik *6*, 235-37 (1964).	R. 219 keV ^{56}Mn $\rightarrow$ Fe
368	P. Ehrhardt, W. Rupp, R. Sizmann: Stopping power of 0.5 to 3.5 MeV α-particles in Cu-Au alloys. Phys. Stat. Sol. *28*, K35-37 (1968).	S. 0.5-3.5 MeV α $\rightarrow$ Cu, Au, Cu-Au alloys
369	S. D. Softky: Ratio of atomic stopping power of graphite and diamond for 1.1 MeV protons. Phys. Rev. *123*, 1085-91 (1961).	S. 1.1 MeV p $\rightarrow$ C

370	S. K. Allison, D. Anton, R. A. Morrison: Stopping power of gases for lithium ions. Phys. Rev. *138*, A688-91 (1965).	S. 0.6-3.75 MeV Li $\to$ H_2, He, CH_4, N_2, CO_2
371	Ya. A. Teplova, I. S. Dimitriev, V. S. Kolaevz, L. N. Fate'eva: On the interaction between lithium ions and matter. J. Exp. Teor. Fiz. *32*, 974-78 (1957) [Engl. trans. Sov. Phys. JETP *5*, 797-800 (1957)].	S, R. 0.5-5 MeV ^7Li $\to$ H_2, Air. R. Same $\to$ Emulsion
372	J. J. Kolats, T. M. Amos, H. Bichsel: Energy-loss straggling of protons in silicon. Phys. Rev. *176*, 484-89 (1968).	δS. 5-42 MeV p $\to$ Si
373	J. R. MacDonald, G. Sidenius: The total ionization in methane of ions with $1 \leq Z_1 \leq 20$ at energies from 10 to 120 keV. Phys. Letters *28A*, 543-44 (1969).	S. 10-120 keV H, He, Li, Be, B, C, N, O, F, Ne, Na, Mg, Al, Si, P, S, Cl, Ar, Ca, V, Sc, Ti $\to$ CH_4
374	H. H. Andersen, H. Simonsen, H. Sørensen: An experimental investigation of charge-dependent deviations from the Bethe stopping power formula. Nuclear Physics *125*, 171-75 (1969).	S. 5-13 MeV p,d; 8-20 MeV ^3He, ^4He $\to$ Al, Ta
375	E. T. Shipatov, B. A. Kononov: Influence of the crystal structure on the loss of energy by fast protons in single crystals of alkali halides. Fiz. Tverd. Tela *10*, 854-57 (1968) [Engl. trans. Sov. Phys. Solid State *10*, 670-72 (1968)].	S, δS. 6.72 MeV p $\to$ NaCl, KBr (cryst.). Chann. and random
376	D. E. Davies: Range and distribution of implanted boron in silicon. Appl. Phys. Letters *13*, 243-45 (1968).	R, δR. 200-400 keV B $\to$ Si
377	R. Kelly: Low-energy depth distributions in Pt, Al and KCl as obtained by sputtering. J. Appl. Phys. *39*, 5298-5303 (1968).	R, δR. 3-9 keV Kr $\to$ Al, Pt, KCl
378	J. L. Whitton, G. Carter, J. H. Freeman, G. A. Gard: The implantation profiles of 10, 20 and 40 keV ^{95}Kr in gallium arsenide. J. Mat. Sci. *4*, 208-17 (1969).	R, δR. 10-40 keV Kr $\to$ GaAs
379	L. Koschmieder: Zur Energiebestimmung von Protonen aus Reichweitemessungen. Z. Naturforschg. *19a*, 1414-16 (1964).	R. 57-144 MeV p $\to$ Cu
380	D. E. Davies: Range of implanted boron, phosphorus, and arsenic in silicon. Can. J. Phys. *47*, 1750-53 (1969).	R, δR. 0.15-1.8 MeV B, 1.0-1.7 MeV As, 0.5-1.7 MeV P $\to$ Si
381	H. Bichsel, C. C. Hanke, J. Buechner: Precision energy loss measurements for natural alpha particles in argon. USC-136-148, p. 1-44 (1969).	S. 1-9 MeV α $\to$ Ar
382	W. K. Chu, D. Powers: Alpha-particle stopping cross sections in solids from 400 keV to 2 MeV. Phys. Rev. *187*, 478-90 (1969).	S. 0.4-2.0 MeV α $\to$ Be, C, Mg, Al, Ti, V, Cr, Mn, Fe, Co, Ni, Cu, Ge, Pd, Ag, In, Sn
383	H. H. Heckman, P. J. Lindstrom: Stopping power differences between positive and negative pions at low velocities. Phys. Rev. Letters *22*, 871-74 (1969).	S. 4-25 MeV π^+, π^- $\to$ Emulsion

384	S. Datz, C. D. Moak, T. S. Noggle, B. R. Appleton, H. O. Lutz: Potential energy and differential-stopping-power function from energy loss spectra of fast ions channeled in gold single crystals. Phys. Rev. *179*, 315-26 (1969).	S. 3 MeV α, 60 MeV ^{127}I $\rightarrow$ Au (cryst.)
385	D. W. Aitken, W. L. Lakin, H. R. Zulliger: Energy loss and straggling in silicon by high-energy electrons, positive pions, and protons. Phys. Rev. *179*, 393-98 (1969).	S, δS. 29-300 MeV p, 50-200 MeV π^+ $\rightarrow$ Si
386	D. Blanchin, J.-C. Poizat, J. Remillieux, A. Sarazin: Experimental determination of the energy loss of protons channeled through aluminum single-crystal. Nucl. Instr. Meth. *70*, 98-102 (1969).	S, δS. 1.4 MeV p $\rightarrow$ Al (cryst.)
387	S. Gorodetzky, A.Pape, E. L. Coopermann, A. Chevallier, J. C. Sens, R. Armbruster: Pouvoir d'Arret du Germanium Pour des Protons d'Ènergie Comprise entre 0.35 et 5.5 MeV. Nucl. Instr. Method *70*, 11-12 (1969).	S. 0.36 - 5.5 MeV p $\rightarrow$ Ge
388	J. J. Ramirez, R. M. Prior, J. B. Swint, A. R. Quinton, R. A. Blue: Energy straggling of alpha particles through gases. Phys. Rev. *179*, 310-14 (1969).	S, δS. 1-3.5 MeV α $\rightarrow$ He, Air, Ar, Kr, Xe
389	W. White, R. M. Mueller: Electron-stopping cross sections of ^{1}H, ^{4}He particles in Cr, Mn, Fe, Co, Ni, and Cu at energies near 100 keV. Phys. Rev. *187*, 499-503 (1969).	S. 25-140 keV H, 40-120 keV He $\rightarrow$ Cr, Mn, Fe, Co, Ni, Cu
390	J. Bøttiger, F. Bason: Energy loss of heavy ions along low-index directions in gold single crystals. Rad. Effects *2*, 105-10 (1969).	S. 300-900 keV N, 320-880 keV Ne, 330-540 keV Na, 370-560 keV Mg, 500-920 keV S, 670-900 keV Cl, 540-970 keV Ar, 340-520 keV K, 370-950 keV Si, 380-660 keV Mn, 430-970 keV Fe, 400-770 keV Kr, 630-940 keV Y, 380-520 keV Mo, 390-760 keV Ag, 490-970 keV Cd, 390-780 keV Sb, 350-760 keV Xe $\rightarrow$ Au (cryst.)
391	G. J. Clark, D. V. Morgan, J. M. Poate: Energy loss of channeled protons in the MeV region, in D. W. Palmer, M. W. Thompson, and P. D. Townsend (edts.): *Atomic Collision Phenomena in Solids*. North-Holland, Amsterdam, p. 388-99 (1970).	S, δS. 4 MeV p $\rightarrow$ CsI, 6-8 MeV p $\rightarrow$ W, 4-6 MeV p $\rightarrow$ Fe, 4-8 MeV p $\rightarrow$ Si, Ge, Mo, NaCl, MgO (all targets cryst.)
392	N. A. Baily, J. E. Steigerwalt: Frequency distribution for very small energy losses by 46 MeV protons. Bull. Am. Phys. Soc. *14*, 846 (1969).	δS. 46 MeV p $\rightarrow$ He-CO_2 mixture
393	A. R. Zander, J. S. Eck, N. R. Fletcher: A simple technique for measuring the stopping power of heavy ions in the few MeV range. Nucl. Instr. Meth. *71*, 343-45 (1969).	S. 2-9 MeV Ca $\rightarrow$ Ca
394	R. Kalish, L. Grodzins, F. Chmara, P. H. Rose: Stopping power of solids for fast moving tantalum ions. Phys. Rev. *183*, 431-35 (1969).	S. 10-140 MeV Ta $\rightarrow$ C, Al, Ag, Au

395	F. Bernhard, U. Müller-Jahreis, G. Rockstroh, S. Schwabe: Stopping cross sections of Li$^+$ ions with energies from 30 to 100 keV in various target materials. Phys. Stat. Sol. *35*, 285-89 (1969).	S. 30-100 keV Li → C, Al, Ti, Ni, Cu
396	Ole Nielsen: Specialeopgave. Niels Bohr Institute, University of Copenhagen, pp. 1-64 (1966).	S, δS. 50 keV ^{12}C → H$_2$, D$_2$, Ne; ^{23}Na, ^{66}Zn, ^{109}Ag, ^{198}Hg, ^{209}Bi → H$_2$, D$_2$, He, Ne, Ar; ^{35}Cl → H$_2$, D$_2$; ^{39}K → D$_2$, He, Ne; ^{55}Mn → Ar; ^{89}Y, ^{178}Hf → N$_2$, D$_2$, Ne, He; ^{175}Lu → D$_2$, He
397	R. Hancock, R. G. Warner, R. L. Woolley: Collision cascades in gold in the energy range 3-169 keV. J. Phys. D: Appl. Phys. *2*, 991-98 (1969).	S. 169 keV Pb → Au
398	E. Leminen, A. Fontell, M. Bister: Stopping power of Al, Zn, and In for 0.6-2.4 MeV protons. Ann. Acad. Sci. Fenn. Ser. A VI. Phys. No. 281 p. 1-12 (1968).	S. 0.6-2.4 MeV p → Al, In, Zn
399	K. Morita, H. Akimura, T. Suita: Energy loss of low energy protons and deuterons in evaporated metallic films. J. Phys.Soc. Jap. *25*, 1525-32 (1968).	S, δS. 7-40 keV p, d → Cu, 7-40 keV p → Be, Al, Ag, Au
400	D. A. Ambrosi, J. L. Wolfson: A method for measuring the energy loss of 92 keV ^{237}Np ions in traversing thin films. Nucl. Instr. Methods. *74*, 251-55 (1969).	S. 92 keV ^{237}Np → Collodium
401	J. W. Hilbert, N. A. Baily, R. G. Lane: Statistical fluctuations of energy deposited in low-atomic-number materials by 43.7-MeV protons. Phys. Rev. *168*, 290-93 (1968).	δS. 43.7 MeV p → He-CO$_2$ mixture
402	E. T. Shipatov: Channeling of high energy protons in ionic single crystals. Fiz. Tved. Tela *10*, 2709-15 (1968). [Engl. trans. Sov. Phys. Solid State *10*, 2132-37 (1969)]	S, δS. 4.7, 6.7 MeV p → NaCl, KCl, KBr (all. cryst.). Random and axial.
403	J. B. Swint, R. M. Prior, J. J. Ramirez: Energy loss of protons in gases. Nucl. Instr. Methods *80*, 134-40 (1970).	S. 0.4-3.4 MeV p → N$_2$, Air, O$_2$, Ne, Ar, Kr, CH$_4$, CO$_2$
404	H. H. Andersen, H. Simonsen, H. Sørensen, P. Vajda: Stopping power of Zr, Gd, and Ta for 5-12 MeV protons and deuterons: Further evidence for an oscillatory behaviour of the excitation potential. Phys. Rev. *186*, 372-75, (1969).	S. 5-12 MeV p,d → Zr, Gd, Ta
405	V. N. Andreev, V. G. Nedopekin, V. I. Rogov: Long-range particles with Z > 2 in ternary fission of U^{235} by thermal neutrons. Yaderna Fiz. *8*, 38-49 (1969). [Eng. trans. Sov. J. Nucl. Phys. *8*, 22-28 (1969)].	S. 0.06-3.0 MeV Li, B; 0.2-10 MeV Ne, 0.1-7 MeV N → Ar

406	P. Hvelplund: Prisopgave. Aarhus University p. 1-105 (1968). (In Danish)	S,δS. 100-200 keV H $\rightarrow$ H$_2$, He; 50-500 keV H $\rightarrow$ Air; 100-500 keV H $\rightarrow$ Ar, Ne; 100-1000 keV H $\rightarrow$ Kr; 100-600 keV He, 100-500 keV Li, 200-500 keV ^{11}B $\rightarrow$ He; 200-400 keV C $\rightarrow$ Ne; 200-500 keV N; 150-500 keV O $\rightarrow$ Mg; 100-300 keV Ag, 100-500 keV Hg $\rightarrow$ H$_2$; 100-400 keV Hg $\rightarrow$ He. δS only. 100-500 keV He, Li, 200-300 keV B, 100-400 keV C, 200-450 keV N, 200-500 keV O, Mg, 200-400 keV Ne $\rightarrow$ Air; 200-400 keV He, 200-300 keV B, C, Ne, 200-500 keV N, O, Mg $\rightarrow$ Ne. S only: 1 MeV p $\rightarrow$ H$_2$, He, Ne, Ar, Kr
407	H. Bätzner: Über die Geschwindigkeitsabnahme von H-Kanalstrahlen in Metallen. Ann. Physik *25*, 233-62 (1936).	S. 4-60 keV p $\rightarrow$ Al, Cu, Ag, Sn, Au
408	M. Mannami, T. Sakurai, K. Ozawa, F. Fujimoto, K. Komaki: Channeling of 1.5 MeV protons in alkali halide crystals. Phys. Stat. Sol. *38*, K1-K4 (1970).	S,δS. 1.5 MeV p $\rightarrow$ NaCl, KCl, KBr, KI (all cryst.)
409	V. N. Andreev, V. G. Nedopekin, V. I. Rogov: Stopping power of argon for ions with Z ranging between 3 and 13. Zh. Eksp. Teor. Fiz. *56*, 1504-07 (1969). [Engl. trans. Sov. Phys. JETP *29*, 807-08 (1969)]	S. 10-20 MeV Li, 8-25 MeV C, N, 4-15 MeV Be, 6-20 MeV B, 10-16 MeV Ne, 20-27 MeV Ar $\rightarrow$ Ar
410	E. P. Arkhipov, Yu. V. Gott: Slowing down of 0.5-30 keV protons in some materials. Zh. Eksp. Teor. Fiz. *56*, 1146-51 (1969). [Engl. trans. Sov. Phys. JETP *29*, 615-18 (1969)]	S. 0.5-30 keV p $\rightarrow$ C, Ti, Al, Cu, Ni, Fe, Ge, Si, Sb, Bi
411	H. Nakata: Ranges of nitrogen ions in Se and energy losses of alpha particles in Al, N, Se, Ag, and Au. Can. J. Phys. *47*, 2545-52 (1969). [Erratum; Can. J. Phys. *48*, 1745 (1970)]	S. 1.4-10 MeV N $\rightarrow$ Se; 1-9 MeV α $\rightarrow$ Al, Ni, Se; 2-5.3 MeV α $\rightarrow$ Ag, Au
412	C. Tschalär, H. D. Maccabee: Energy-straggling measurements of heavy charged particles in thick absorbers. Phys. Rev. B*l*, 2863-69 (1970).	δS. 20, 49 MeV p, 80 MeV α $\rightarrow$ Al, Au
413	E. S. Machlin, S. Petralia, A. Desalvo, R. Rosa, F. Zignani: Energy loss of protons channeled through very thin gold. Phil. Mag. *22*, 101-16 (1970).	S,δS. 92 keV p $\rightarrow$ Au (cryst.)
414	G. J. Igo, D. D. Clark, R. M. Eisberg: Statistical fluctuations in ionization by 31.5 MeV protons. Phys. Rev. *89*, 879-80 (1953).	δS. 31.5 MeV p $\rightarrow$ NaI
415	T. Andersen, G. Sørensen: A sectioning technique for copper, silver, and gold and its application to penetration and diffusion studies. Rad. Effects *2*, 111-17 (1969).	R,δR. 60 keV ^{67}Cu, ^{57}Co, ^{32}P $\rightarrow$ Cu, 100-400 keV ^{85}Kr $\rightarrow$ Ag, 30-60 keV ^{85}Kr $\rightarrow$ Au

416	F.Fehsenfeld, A. Scharmann: Messungen der Eindringtiefen von Ionen in LiF-ZnS-und CsJ-Aufdampfschichten. Z. Physik *230*, 435-42 (1970).	R. 5-60 keV H, He Ne, Ar, Kr → LiF, ZnS, CsJ
417	H. Bach: Zur Bestimmung der Reichweiten von beschleunigten Ionen in dünner Oxidschichten. Z. angew. Phys. *28*, 239-44 (1970).	R. 4.2-5.6 keV Ar → SiO_2, TiO_2
418	M. Derrick, T. Fields, L. G. Hyman, G. Keyes, J. Fetkovich, J. McKenzie, I. T. Wang: Range-energy relation in helium. Phys. Rev. A*2*, 7-13 (1970).	R. 30.6 MeV ^{3}H, 8.43 MeV ^{4}H, 4.12 MeV μ^+, μ^- → He
419	J. Mory, D. de Guilebon, G. Delsarte: Mesure du Parcours Moyen des Fragments de Fission avec Le Mica comme Detecteur-Influence de la Texture Cristalline. Rad. Effects *5*, 37-40 (1970).	R. Fiss. Fragm. → Al, Ti, Fe, Ni, Cu, Zr, Nb, Mo, Pd, Ag, Ta, W, Au
420	J. C. Majure, J. W. Hooper: An experimental investigation of the characteristic energy losses of 3-10 keV lithium particles in thin films. Georgia Inst. Tech. Rep. ORO-3027-20, pp. 1-179 (1971).	S,δS. Dep. on scatt. angle. 3-10 keV Li → C
421	P. Hvelplund: Energy loss and straggling of 100-500 keV atoms with $2 \le Z_1 \le 12$ in various gases. Kgl. Danske Videnskab. Selskab Mat. Fys. Medd. *38*, No. 4, p. 1-25 (1971).	S,δS. 100-500 keV Li, 200-500 keV Be, B, C, N, O, F, Ne, Na, Mg → Air, He, Ne; 100-300 keV He, O → H_2, O_2
422	R. Hellborg: The energy loss of channeled protons determined in an indirect way. Phys. Scripta *4*, 75-82 (1972).	S. 1.4-1.8 MeV p → BaF_2, CaF_2 (both cryst.)
423	R. L. Lander, W. A. Mehlhop, H. J. Lubatti, G. L. Schnurmacher: Solid-state devices as detectors of high-energy interactions. Nucl. Instr. Methods *42*, 261-68 (1966).	δS. 760 MeV p → Si
424	T. T Hoang, R. Drouin: Aluminum oxide films produced by a two step proces and alpha stopping power studies. Nucl. Instr. Methods *87*, 297-98 (1970).	S. 5.5-8.8 MeV α → Al_2O_3
425	N. A. Baily, J. E. Steigerwalt, J. W. Hilbert: Frequency distrbution of energy deposition by fast charged particles in very small pathlengths. Phys. Rev. B*2*, 577-82 (1970).	δS. 46.4 MeV p → He-CO_2 mixture
426	G. Högberg, H. Nordén, R. Skoog: Energy loss and energy straggling of well channelled hydrogen, helium and lithium ions in gold. Phys. Stat. Sol. *42*, 441-51 (1970).	S,δS. 2-54 keV H, D, He, Li → Au (crtst.)
427	L. E. Porter, L. C. McIntyre, W. Haeberli: Stopping power of havar for 2.5 to 7.0 MeV deuterons. Nucl. Instr. Methods *89*, 237-43 (1970).	S. 2.5-7.0 MeV d → Havar
428	R. L. Hines: Relative energy losses of O^+ ions channeled in gold. Phys. Lettes *33A*, 348-49 (1970).	S,δS. 20-28 keV d → Au (cryst.)
429	E. Bonderup, P. Hvelplund: Stopping power and energy straggling of swift protons. Phys. Rev. A*4*, 562-69 (1971).	S,δS. 100-500 keV p → H_2, He, Air, Ne, Ar, Kr

430	A. Johansen, S. Steenstrup, T. Wohlenberg: Energy loss of protons in thin films of carbon aluminum and silver. Rad. Effects *8*, 31-32 (1971).	S. 70-90 keV p → C, Al, Ag
431	U. Hoyer, H. Wäffler: Der atomare Bremsquerschnitt von H_2, D_2, He, N_2 und A für α-Teilchen in Umladungsgebiet. ($0.5 \leq E_\alpha \leq 2$ MeV). Z. Naturforschg. *26a*, 592-95 (1971).	S. 0.5-2.0 MeV α → H_2, D_2, N_2, Ar
432	M. Hakim, N. H. Schafrir: ^{252}Cf fission fragment energy loss measurements in elementary gases and solids as compared with theory. Can. J. Phys. *49*, 3024-35 (1971).	S. Fiss. fragm. → H_2 D_2, He, C, N_2 O_2, Ne, Al, Ar, Ni, Cu, Kr, Ag, Xe, Au
433	A. van Wijngaarden, B. Miremadi, W. E. Baylis: Energy spectra of keV backscattered protons as a probe for surface region studies. Can. J. Phys. *49*, 2440-48 (1971).	S. 20-100 keV H, He → Au
434	D. Ward, R. L. Graham, J. S. Geiger: Measurement of stopping power for ^{4}He, ^{16}O and ^{35}Cl ions at ~1 to ~3 MeV per nucleon in Ni, Ge, Y, Ag, and Au. Can. J. Phys. *50*, 2302-12 (1972).	S. 3-15 MeV α, 8-66 MeV O, 10-90 MeV ^{35}Cl → Ni, Ge, Y, Ag, Au
435	R. Ishiwari, N. Shiomi, S. Shirai, T. Ohata, Y. Uemura: Comparison of stopping powers of Al, Ni, Cu, Rh, Ag, Pt and Au for protons and deuterons of exactly the same velocity. Bull. Inst. Chem. Res. Kyoto Univ. *49*, 390-402 (1971).	S. 7.2 MeV p, 14.4 MeV d → Al, Ni, Cu, Rh, Ag, Pt, Au
436	Idem: Stopping power of Be, Al, Cu, Mo, Ta and Au for 28 MeV Alpha Particles. Bull. Inst. Chem. Res. Kyoto Univ. *49*, 403-08 (1971).	S. 28 MeV α → Be, Al, Cu, Mo, Ta, Au
437	E. Rotondi: Energy loss of alpha particles in tissue. Rad. Research *33*, 1-9 (1968).	S. 0.1-5.3 MeV α → N_2, O_2, CH_4, CO_2
438	E. Rotondi: Bragg's additivity law of stopping power for 5 MeV α particles in O_2, N_2, CO_2, CO, NH_3 and hydrocarbon gases. NRC Canada Report No. NRC-9076 p. 1-6 (1966).	S. 5 MeV α → N_2, O_2, CO, CO_2, NH_3, Hydrocarbons
439	P. D. Bourland, W. K. Chu, D. Powers: Stopping cross section of gases for alpha particles from 0.3 to 2.0 MeV. Phys. Rev. B*3*, 3625-35 (1971).	S. 0.3-2.0 MeV α → H_2, O_2, N_2, NH_3, N_2O, CO, CO_2, CH_4, C_2H_2, C_2H_4, C_2H_6, C_3H_6, $(CH_3)_2$
440	P. D. Bourland, D. Powers: Bragg-rule applicability to stopping cross sections of gases for alpha particles of energy 0.3-2.0 MeV. Phys. Rev. B*3*, 3635-41 (1971).	S. 0.3-2.0 MeV α → Targets from ref. 439
441	J. A. Penkrot, B. L. Cohen, G. R. Rao, R. H. Fulnier: Energy loss straggling of protons in nickel. Nucl. Instr. Methods *96*, 505-08 (1971).	δS. 17 MeV p → Ni
442	J. H. Ormrod: Electronic stopping cross sections of deuterons in titanium. Nucl. Instr. Methods *95*, 49-51 (1971).	S. 30-100 keV p, d → Ti
443	R. Ishiwari, N. Shiomi, S. Shirai, Y. Uemara: Stopping powers of Al, Ti, Fe, Cu, Mo, Ag, Sn and Au for 7.2 MeV protons. Bull. Inst. Chem. Res. Kyoto Univ. *52*, 19-39 (1974).	S. 7.2 MeV p → Al, Ti, Fe, Cu, Mo, Ag, Sn, Ta, Au

444	H. R. Rubin, R. A. Burnstein, R. C. Misra: Low-velocity moderation of Σ^- hyperons in hydrogen. Phys. Rev. A3, 1427-34 (1971).	R. 0.1-8.0 MeV $\Sigma^- \rightarrow$ Emulsion
445	P. H. Garbincius, L. G. Hyman: Range-energy relation in hydrogen. Phys. Rev. A2, 1834-38 (1970).	R. 12-40 MeV p $\rightarrow$ H_2
446	J. Leon, N. H. Steiger-Shafrir: Range and range straggling of 97 keV ^{224}Ra particles in gases. Can. J. Phys. 49, 1004-17 (1971).	R, δR. 97 keV ^{224}Ra $\rightarrow$ H_2, He, N_2, O_2, Ne, Ar, Kr, Xe
447	J. Leon, N. H. Steiger-Shafrir: Investigation of nuclear stopping mechanism by heavy particle range measurements in isotopic media. Can. J. Phys. 49, 2106-17 (1971).	R. 97 keV ^{224}Ra $\rightarrow$ H_2, D_2, T_2, ^{3}He, ^{4}He, $^{14}N_2$, $^{15}N_2$, $^{16}O_2$, $^{18}O_2$, ^{20}Ne, ^{22}Ne
448	W. Schimmerling, K. G. Vosburgh, P. W. Todd: Measurements of range in matter for relativistic heavy ions. Phys. Rev. B7, 2895-99 (1973).	R. 50-270 MeV N, 258 MeV Ne $\rightarrow$ Polyethylene; 44-270 MeV N, 284 MeV Ne, Ar $\rightarrow$ Polymethylacrylat; 270 MeV N $\rightarrow$ Al, Cu, Pb
449	J. Comas, W. Lucke, A. Addamiano: Ion implanted Al concentration profiles in unannealed and annealed 6H-SiC. Bull. Am. Phys. Soc. 18, 606 (1973).	R, δR. 60 keV Al $\rightarrow$ SiC
450	K. L. Dunning, J. Comas, G. K. Hubler: Depth profiles of aluminum implanted into SiC. Nuclear resonance method. Bull. Am. Phys. Soc. 18, 606 (1973).	R,δR. 60 keV Al $\rightarrow$ SiC
451	R. A. Langley: Range-energy relations for N, Na, and Ar ions (0.3-2.0 MeV) in Ar, N_2, O_2, and air. Phys. Rev. A6, 1863-69 (1972).	R. 0.3-2.0 MeV N, Na $\rightarrow$ Air; 0.3-1.0 MeV Ar $\rightarrow$ Air, N_2, O_2, Ar
452	V. G. Reddi, J. D. Sanbury: Channeling of phosphorus ions in silicon. Appl. Phys. Lettes, 20, 30-31 (1972).	R,δR. 30-600 keV ^{31}P $\rightarrow$ Si (cyrst.)
453	J. F. Ziegler, B. L. Crowder, G. W. Cole, J. E. E. Baglin, B. J. Masters: Boron atom distributions in ion-implanted silicon by the (n,^{4}He) nuclear reaction. Appl. Phys. Letters 21, 16-18 (1972).	R,δR. 40-500 keV B $\rightarrow$ Si
454	B. L. Crowder: The influence of the amorphous phase on ion distributions and annealing behavior of group III and group V ions implanted into silicon. J. Electrochem. Soc. 118, 943-52 (1971).	R,δR. 50-200 keV B, 200 keV Al; 140, 280 keV ^{69}Ga, ^{71}Ga; 100-280 keV ^{31}P; 280 keV ^{75}As; 120, 260 keV ^{121}Sb, ^{123}Sb; 240 keV ^{209}Bi $\rightarrow$ Si
455	R. A. Moline, G. W. Reutlinger: Phosphorus channeled in silicon: Profiles and electrical activity; in I. Ruge and J. Graul: *Ion Implantation in Semiconductors*. Springer, Berlin. pp. 58-69 (1971).	R,δR. 100-300 keV P $\rightarrow$ Si (cryst.)
456	T. E. Seidel: Distribution of boron implanted into silicon. Idem. p. 47-57 (1971).	R,δR. 30-300 keV ^{11}B $\rightarrow$ Si (cryst. and polycryst.)

457	R. A. Langley: Range-energy relations for helium ions and protons in Ar, N_2, O_2, and air (0.2-2.0 MeV). Phys. Rev. A *4*, 1868-72 (1971).	R. 0.2-2.0 MeV p, $\alpha \rightarrow N_2$, O_2, Ar, air
458	W. R. Pierson, J. T. Kummer, W. Brachuczek: Ranges of recoil atoms from the (n,γ) proces. Phys. Rev. B*4*, 2846-53 (1971).	R. ~ 50 eV Au $\rightarrow$ D, He, Ne, Ar, Xe
459	P. Sebilotte, M. Badanoin, V. B. Ndocko, V. Siffert: Low energy boron implantation profiles in silicon from junction depth measurements. Rad. Effects *7*, 7-15 (1972).	R,δR. 15 keV B $\rightarrow$ Si
460	T. Andersen, A. Ebbesen: A sectioning technique for sodium chloride single crystals and its application for ion implantation studies. Rad. Effects. *11*, 113-18 (1971).	R,δR. 30-60 keV Kr, 40 keV P $\rightarrow$ NaCl (cryst.)
461	M. Bister, A. Anttila, A. Fontell, E. Leminen: A method for the determination of recoil ion ranges needed in DSA measurements. 2. Physik *250*, 82-86 (1972).	R. 50 keV Al $\rightarrow$ C, Cu, Mo, Ta
462	A. R. Clark, R. C. Field, H. J. Frisch, W. R. Holley, R. P. Johnson, L. T. Kerth, R. C. Sah, W. A. Wenzel: Observed difference in the ranges of positive and negative muons. Phys. Lettes *41B*, 229-33 (1972).	R. 850-1100 MeV/c μ^+, μ^- $\rightarrow$ Fe
463	G. della Mea, A. V. Drigo, S. lo Russo, P. Mazzoldi: Indirect determination of the energy loss of protons channeled in silicon crystals. Rendiconti della Academia Nationale dei Lincei. Classe di Scienze fisiche matematiche e naturali. Ser. 8, Vol. *52*, No. 5, p. 727-33 (1972).	S. 1600 keV p $\rightarrow$ Si (cryst.)
464	A. K. M. M. Haque, R. M. Hora: Energy loss and straggling of alpha particles in argon. Nucl. Instr. Methods *104*, 77-83 (1972).	S,δS. 1-8 MeV $\alpha \rightarrow$ Ar
465	J. A. Borders: Helium ion stopping cross sections in gold. Rad. Effects. *16*, 253-57 (1972).	S. 0.4-1.9 MeV $\alpha \rightarrow$ Au
466	C. Foster, W. H. Kool, W. F. van der Weg, H. E. Roosendaal: Random stopping power for protons in silicon. Rad. Effects *16*, 139-40 (1972).	S. 120 keV p $\rightarrow$ Si
467	G. T. Huetter, R. Madey, S. M. Yushak: Fluctuations in the energy loss of 66- and 100-MeV protons in a thin proportional counter. Phys. Rev. A*6*, 250-55 (1972).	δS. 66, 100 MeV p $\rightarrow$ (0.9 Xe, 0.1 CH_4)
468	D. Powers, W. K. Chu, R. J. Robinson, A. S. Lodhi: Measurement of molecular stopping cross sections of halogencarbon compounds and calculation of atomic cross sections of halogens. Phys. Rev. A*6*, 1425-35 (1972).	S. 0.3-2.0 MeV $\alpha \rightarrow CF_4$, C_2F_6, C_3F_8, CCl_4, $CClF_3$, CCl_2F_2, $CHCl_2$, $CBrF_3$, C_2H_3Br, C_2H_5Br, C_2H_5I
469	H. J. Hirsch, Hj. Matzke: Stopping power and range of α-particles in (U,Pu)C and UC and application to selfdiffusion measurements using alphaspectroscopy. J. Nucl. Mater. *45*, 29-39 (1972).	S,δS. 1-5 MeV $\alpha \rightarrow$ U, UC, (U,Pu)C

470	V. Nitzki, Hj. Matzke: Stopping power of 1 to 9 MeV He^{++}-ions in UO_2, $(U,Pu)O_2$, and ThO_2. Phys. Rev. B*8*, 1894-1900 (1973).	S. 1-9 MeV $\alpha \rightarrow UO_2$, ThO_2, $(U,Pu)O_2$
471	H. Nann, W. Schäfer: The energy straggling of protons in aluminium. Nucl. Instr. Methods *100*, 217-19 (1972).	δS. 8-19 MeV p $\rightarrow$ Al
472	M. B. Al-Bedri, S. J. Harris, D. A. Sykes: The dependence of energy straggling on the atomic number of absorber. Nucl. Instr. Methods *106*, 241-43 (1973).	δS. 5.2 MeV $\alpha \rightarrow$ He, Ne, Ar, Kr, CH_4, CO_2, N_2, Air.
473	I. A. Abroyan, V. A. Koryukin: Retardation of protons in chromium and copper. Fiz. Tverd. Tela *13*, 3112-14 (1971). [Engl. trans. Sov. Phys. Solid State *13*, 2614-16 (1972)].	S. 0.6-10 keV p $\rightarrow$ Cu, Cr
474	J. F. Ziegler, M. H. Brodsky: Specific energy loss of ^{4}He ions in silicon (amorphous, polycrystalline, and single crystal). J. Appl. Phys. *44*, 188-96 (1973).	S. 0.42-2.75 MeV $\alpha \rightarrow$ Si
475	H. Nakata: Analysis of energy-loss data for 0.2-5.0 MeV/amu p,α and N in Se. Phys. Rev. B*3*, 2847-51 (1971).	S. 0.7-1.4 MeV p $\rightarrow$ Al, Se, Ag
476	L. Bridwell, A. L. Walters, Jr.: Stopping cross sections for fission fragments of ^{252}Cf by gold, silver, and carbon. Phys. Rev. B*3*, 2149-53 (1971).	S. Fiss. fragm. $\rightarrow$ C, Ag, Au
477	M. D. Brown, C. D. Moak: Stopping powers of some solids for 30-90-MeV ^{238}U ions. Phys. Rev. B*6*, 90-94 (1972).	S. 30-90 MeV ^{238}U $\rightarrow$ C, Al, Ni, Ag, Au
478	A. Valenzuela, W. Meckbach, A. J. Kestelman, J. C. Eckardt: Stopping power of some pure metals for 25-250-keV hydrogen ions. Phys. Rev. B*6*, 95-102 (1972).	S rel. to 250 keV p. 25-250 keV p $\rightarrow$ Ni, Cu, Ag, Sn, Au.
479	G. Högberg: Electronic and nuclear stopping cross sections in carbon. Phys. Stat. Sol. (b), *48*, 829-41 (1971).	S. 10-46 keV ^{7}Li, 15-46 keV ^{11}B, ^{14}N, 22-46 keV ^{12}C, ^{18}O, ^{19}F, ^{20}Ne, ^{22}Na, 32-46 keV ^{31}P, ^{40}Ar $\rightarrow$ C
480	G. della Mea, A. V. Drigo, S. lo Russo, P. Mazzoldi, G. G. Bentini: Transmission energy loss of light channeled particles in thin silicon single crystals. Rad. Effects *13*, 115-19 (1972).	S,δS. 0.9-5.0 MeV p, d, $\alpha \rightarrow$ Si (cryst.)
481	K. Björkquist, B. Domeij: Stopping power of C, N, and O ions in Cr, Fe, Co, Ni, Cu, and Zn in the 1 MeV region. Rad. Effects *13*, 191-96 (1972).	S. 0.5-2.0 MeV C, O, N $\rightarrow$ Cr, Fe, Co, Ni, Cu, Zn
482	F. H. Eisen, G. J. Clark, J. Bøttiger, J. M. Poate: Stopping power of energetic helium ions transmitted through thin silicon crystals in channeling and random directions. Rad. Effects *13*, 93-100 (1972).	S,δS. 0.1-18 MeV $\alpha \rightarrow$ Si (cryst.). Chan. and random.
483	B. R. Appleton, J. H. Barrett, T. S. Noggle, C. D. Moak: Orientation dependence of intensity and energy loss of hyperchanneled ions. Rad. Effects, *13*, 171-81 (1972).	S,δS. 21.6-60 MeV ^{127}I, 3 MeV $\alpha \rightarrow$ Au, Ag (both cryst.)

484	B. Sellers, A. Hanser, J. G. Kelley: Energy loss and stopping power measurements between 2 and 10 MeV/.amu for ^{3}He and ^{4}He in silicon. Phys. Rev. B8, 98-102 (1973).	S. 6-30 MeV ^{3}He, 8-40 MeV ^{4}He → Si (cryst.?)
485	J. G. Kelley, B. Sellers, F. A. Hanser: Energy-loss and stopping power measurements between 2 and 10 MeV/amu for ^{12}C, ^{14}N, and ^{16}O in silicon. Phys. Rev. B8, 103-06 (1973).	S. 24-120 MeV C, 28-140 MeV N, 32-160 MeV O → Si (cryst?)
486	E. I. Sirotinen, A. F. Tulinov, A. Fiderkevich, K. S. Shyshkin: The determination of energy losses from the spectrum of particles scattered by a thick target. Rad. Effects. 15, 149-52 (1972).	S rel. to 6 MeV p → Pb. 1-25 MeV α → W; 1-6 MeV p → Pb, Ta, Mo, W, Ag, Yb, Ce.
487	G. Högberg: Stopping cross section for 50 keV neon ions scattered in thin carbon films. Phys. Letters $35A$, 327-28 (1971).	S. Dep. on scatt. angle, 50 keV Ne → C
488	G. Högberg, R. Skoog: Non-evidence for Z_1, oscillations of the nuclear ion-atom interaction in an amorphous target. Rad. Effects 13, 197-202 (1972).	S. 50 keV ^{7}Li, ^{11}B, ^{12}C, ^{14}N, ^{16}O, ^{19}F, ^{20}Ne, ^{23}Na, ^{24}Mg, ^{31}P, ^{40}Ar → C
489	K. C. Shane, G. G. Seaman: Energy loss of ^{20}Ne ions in aluminum. Phys. Rev. B8, 86-89 (1973).	S. 18.5, 19.8 MeV ^{20}Ne → Au
490	E. Leminen, A. Anttila: Energy loss and straggling of 0.6-2.0 MeV protons in Fe, Co and Sb. Ann. Acad. Sci. Fenn. Ser. A VI, Physics, No. 370 p. 1-15 (1971).	S. 0.6-2.0 MeV p → Fe, Co, Sb
491	G. L. Cano: Penetration of low-energy protons through thin films. J. Appl. Phys. 43, 1504-07 (1972).	S. 10-30 keV p → Er$_2$O$_3$, Sc$_2$O$_3$, Au
492	D. A. Thompson, W. D. Mackintosh: Stopping cross sections for 0.3 to 1.7-MeV helium ions in silicon and silicion dioxide. J. Appl. Phys. 42, 3969-76 (1971).	S. 0.3-l.7 MeV α → Si, SiO$_2$
493	E. Leminen: Stopping power of Ti, Mo, Ta, and W for 0.50 to 1.75 MeV protons. Ann. Acad. Sci. Fenn. Ser. A VI, Phys. No. 386 p. 1-14 (1972).	S. 0.5-1.75 MeV p → Ti, Mo, Ta, W
494	T. Jokic, B. Perovic: Xenon ion ranges in copper single crystals. Phys. Stat. Sol. (a) 18, K 73-76 (1973).	R,δR. 20-100 keV Xe → Cu (cryst.)
495	T. Jokic: Xenon ion ranges in polycrystalline gold. Phys. Stat. Sol. (a) 18, K 77-80 (1973).	R,δR. 40-80 keV Xe → Au
496	G. Eldridge, F. Chernow, G. Rise: Further studies of bismuth-implanted cadumium sulfide. J. Appl. Phys. 44, 3858-61 (1973).	R,δR. 25 keV ^{209}Bi → CdS (cryst.)
497	W. K. Lin, H. G. Olson, D. Powers: Alpha-particle stopping cross sections of silicon and germanium. J. Appl. Phys. 44, 3631-34 (1973).	S. 0.3-2.0 MeV α → Si, Ge (amorph.)
498	V. G. K. Reddi, J. D. Sansbury: Channeling and dechanneling of ion-implanted phosphorus in silicon. J. Appl. Phys. 44, 2951-63 (1973).	R,δR. 30-900 keV ^{31}P → Si (cryst.)

499	H. Sørensen, H. H. Andersen: Stopping power of Al, Cu, Ag, Au, Pb and U for 5-18-MeV protons and deuterons. Phys. Rev. B8, 1854-63 (1973).	S. 5-18 MeV p, d → Al, Cu, Ag, Au, Pb, U
500	W. K. Lin, H. G. Olson, D. Powers: Alpha-particle stopping cross section of solids from 0.3 to 2.0 MeV. Phys. Rev. B8, 1881-88 (1973).	S. 0.3-2.0 MeV α → Se, Y, Zr, Nb, Mo, Sb, Te, La, Dy, Ta, W, Au
501	G. Linker, O. Meyer, M. Gettings: Back-scattering energy loss parameters measurements in thin metal films. Thin Solid Films 19, 177-85 (1973).	S. 2 MeV α → Ni, V, Ni, Mo, Ta
502	R. A. Langley, R. S. Blewer: Technique for accurately measuring stopping cross sections of reactive metals by ion backscattering: He ions in erbium. Thin Solid Films 19, 187-94 (1973).	S. 0.25-2.5 MeV He → Er
503	J. S.-Y. Feng, W. K. Chu, M.-A. Nicolet, J. W. Mayer: Relative measurements of stopping cross section factors by backscattering. Thin Solid Films 19, 195-204 (1973).	S relative. 1-2 MeV α → Au rel. to Ag, Cu rel. to Ag, Au rel. to Cu, Au rel. to Al, Al rel. to Si.
504	D. Powers, A. S. Lodhi, W. K. Lin, H. L. Cox, Jr: Molecular effects in the energy loss of alpha particles in gasous media. Thin Solid Films, 19, 205-15 (1973).	S. 0.3-2.0 MeV α → CO, CO_2, C_2H_3Br, C_2H_5Br, $CBrF_3$, $C_2Br_2F_4$, $(CH_3)_2O$, $C_2H_2F_2$, Hydrocarbons.
505	O. Meyer, G. Linker, B. Kraeft: Validity of Bragg's rule in sputtered superconducting NbN and NbC films of various compositions. Thin Solid Films. 19, 217-26 (1973).	S. 2 MeV α → NbN, NbC
506	J. S.-Y. Feng, W. K. Chu, M-A. Nicolet: Bragg's rule study in binary metal alloys and metal oxides for MeV $^4He^+$ ions. Thin Solid Films 19, 227-36 (1973).	S. 0.5-2.25 MeV α → AuAg, AuCu, AuAl, Fe_2O_3, Fe_3O_4, Al_2O_3
507	J. Bøttiger, F. H. Eisen: On conversion from an energy scale to a depth scale in channeling experiments. Thin Solid Films 19, 239-46 (1973).	S. 0.2-0.4 MeV p → Si (cryst.)
508	R. Behrisch, B. M. U. Schertzer: Rutherford backscattering as a tool to determine electronic stopping powers in solids. Thin Solid Films 19, 247-57 (1973).	S. 50-150 keV p → Nb, Ta, Ta_2O_5
509	J. M. Harris, W. K. Chu, M.-A. Nicolet: Energy straggling of ^4He below 2 MeV in Pt. Thin Solid Films 19, 259-65 (1973).	S,δS. 1-2 MeV α → Pt
510	S. Furukawa, H. Matsumura, H. Ishiwara: Back-scattering study of heavy-ion distribution in semiconductors. Thin Solid Films 19, 399-406 (1973).	R,δR,$\delta R_\perp$. 180 keV Kr → Si
511	R. Naidu: Courbes d'ionisation dans les krypton et le xénon purs relatives aux rayons α du polonium. J. Phys. Radium 5, 343-46 (1934).	R. 5.3 MeV α → Kr, Xe
512	R. Naidu: Etude des courbes d'ionisation des rayons α. Ann. de Physique 1, 72-122 (1934).	R. 5.3, 7.7 MeV α → Air

513	R. I. Ewing: Response of silicon surface barrier detectors to hydrogen ions of energies 25 to 250 keV. IRE Trans. Nucl. Sci. NS9, No. 3, 207-10 (1962).	S. rel. to H^+; 70 keV/nucleon H^+_2, H^+_3 → Si
514	Y. U. Galaktionov, F. A. Yech, V. Z. Lyubimov: Investigations of fluctuations of measurements of particles ionization power in a spark chamber. Nucl. Instr. Methods 33, 353-54, (1965).	δS. 600 MeV/c p, π → Ne
515	G. W. Grew: Cyclotron tests to determine the response of solid-state detectors to protons of energies 50 to 160 MeV for use in a proton spectrometer. IEEE Trans. Nucl. Sci. NS 12, 308-13 (1965).	S,δS. 50-160 MeV p → Si
516	R. Schuch: Blocking-Effekte bei Transmission von α-Teilchen durch Germanium - und Siliziumkristalle. Z. Physik A 272, 61-66 (1975).	S. 8.8 MeV α → Si, Ge (cryst.)
517	V. Dose, G. Sele: Die elektronische Bremsvermögen von Stickstoff und Sauerstoff für niederenergetische Protonen. Z. Physik A 272, 237-43 (1975).	S. 7-30 keV p → O_2, N_2
518	R. Schimko, C. E. Richter, K. Rogge, G. Schwartz, M. Trapp: Implanted arsenic and boron concentration profiles in SiO_2 layers. Phys. Stat. Sol. (a) 28, 87-93 (1975).	R,δR. 40-300 keV ^{11}B, 40-150 keV ^{75}As, → SiO_2
519	K. Schmid, G. Fischer, H. Müller, H. Ryssel: Experimental data about dechanneling and channel stopping power. Rad. Effects 23, 145-49 (1974).	S. 1 MeV α → Si (cryst.)
520	W. K. Hofker, D. P. Oosthoek, N. J. Koeman, H. A. M. de Grefte: Concentration profiles of boron implantations in amorphous and polycrystalline silicon. Rad. Effects 24, 223-31 (1975).	R,δR. 30-200 keV B → Si (polycryst.), 70-800 keV B → Si (amorph.)
521	J. M. Harris, M.-A. Nicolet: Energy straggling of ^{4}He ions below 2 MeV in Al, Ni, Pt, and Au. J. Vac. Sci. Technol. 12, 439-43 (1975).	S,δS. 0.6-2.0 MeV α → Al, Ni, Pt, Au
522	G. Betz, H.-J. Isele, E. Rössle, G. Hortig: Stopping powers of He, air, and Kr for 5 to 110 MeV uranium ions. Nucl. Instr. Methods. 123, 83-87 (1975).	S. 5-110 MeV U → He, Air, Kr
523	J. C. Duder, J. F. Clare, N. Naylor: Stopping power of Havar for 0.8-3.9 MeV deuterons and 2.9-6.0 MeV protons. Nucl. Instr. Methods 123, 89-91 (1975).	S. 0.8-3.9 MeV d, 2.9-6.0 MeV p → Havar (mainly Co).
524	V. Martini: Energy-loss measurements of keV-ions in gases by time-of-flight spectroscopy. Nucl. Instr. Methods 124, 119-24 (1975).	S. 25-75 keV Pb → He
525	E. Liukkonen, V. Metag, G. Sletten: Recoil escape efficiencies and ranges of actinide recoils from actinide targets. Nucl. Instr. Methods 125, 113-17 (1975).	R. 60-400 keV ^{240}Cm, → ^{241}AmO$_2$
526	V. W. E. Burgess: The stopping power of organic compounds for alpha particles in the range 1.5-8 MeV. J. Phys. D: Appl. Phys. 8, 782-89 (1975).	S. 1.5-8 MeV α → organic compounds (sol., liq., vap.)

527	M. Iwaki, K. Gamo, K. Masuda, S. Namba, S. Ishihara, I. Kimura, K. Yohota: Comparison between concentration profiles of arsenic implanted in silicon measured by means of neutron activation analysis and radioactive ion implantation. Jap. J. Appl. Phys. *14*, 167-68 (1975).	R. 45 kev As → Si
528	W. Lucke, J. Comas, G. Hubler, K. Dunning: Effect of annealing on profiles of aluminum implanted into silicon carbide. J. Appl. Phys. *46*, 994-97 (1975).	R. 60 keV Al → SiC
529	S. Myers, R. A. Langley: Study of the diffusion of Au and Ag in Be using ion beams. J. Appl. Phys. *46*, 1034-42 (1975).	R. 100 keV Au → Be
530	R. P. Lyons, Jr., J. E. Ehret, Y. S. Park: Ion implantation of diatomic sulpher into GaAs. Bull. Am. Phys. Soc. *20*, 318 (1975).	R. ~ 100 keV S_2 → GaAs.
531	P. Scholten, M. Skokan, K. W. Kemper, W. G. Moulton: Range and distributin of Gd implanted in Nb as determined by ^{16}O backscatter method. Bull. Am. Phys. Soc. *19*, 1117 (1974).	R. 50-150 keV Gd → Nb
532	W. M. Gibson, R. Laubert, H. E. Wegner: Energy loss of O⁻ and O⁻$_2$ beams in thin carbon and nickel foils. Bull. Am. Phys. Soc. *20*, 619 (1975).	S. 2.9 MeV O⁻, 5.8MeV O$_2$⁻ → C, Ni
533	J. W. Tape, W. M. Gibson, J. Remillieux: The energy loss of H⁺ and H⁺$_2$ beams in thin carbon foils. Bull. Am. Phys. Soc. *20*, 618 (1975).	S. 1 MeV H⁺, 2 MeV H⁺$_2$ → C
534	H. Faraggi: Détermination expérimentale des relation parcours-énergie et du pouvoir de ralentissement des émulsions nucléaires pour les particles chargées de faible énergie. C. R. Acad. Sci. *230*, 1398-99 (1950).	R. 0.5-4.0 MeV α → Emulsion
535	T. S. Gooding, R. M. Eisberg: Statistical fluctuations of energy loss of 37-MeV protons. Phys. Rev. *105*, 357-60 (1957).	δS. 37 MeV p → Ar, plast. scint.
536	S. M. Myers, W. Beetzhold, S. T. Picraux: Implantation and diffusion of Cu in Be, in B. L. Crowder (edt): *Ion Implantation in Semiconductors and Other Materials*. Plenum. N. Y. p. 445-64 (1973).	R,δR. 100 keV Cu → Be
537	J. L. Combasson, J. Bernard, G. Guernet: Physical profile measurements in insulating layers using the ion analyzer. Idem. p. 285-94 (1973).	R,δR. 60, 100 keV B → SiO_2; 20, 40 keV B → Si_3N_4; 60 keV B → Si (amorphous)
538	B. L. Crowder, J. F. Ziegler, G. W. Cole: The influence of the amorphous phase on boron atom distributions in ion implanted silicon. Idem. p. 257-66 (1973).	R,δR. 50-150 keV ^{10}B → Si. (cryst. and amorph.)

539	W. K. Chu, B. L. Crowder, J. W. Mayer, J. F. Ziegler: Ranges and distributions of ions implanted into dielectrics. Idem. p. 225-41 (1973).	R.δR. 140, 280 keV Zn, 280 keV Ga, 300 keV As, 150-280 keV Se, 260 keV Cd, 280 keV Te $\rightarrow$ SiO_2; 200 keV Se, 260 keV Ga, Cd, Te, 280 keV Zn $\rightarrow$ Si_3N_4; 260 keV Zn, Ga, As, Se, Cd, Te $\rightarrow$ Al_2O_3
540	S. Furukawa, H. Matsumura: Theoretical and experimental studies of lateral spread of implanted ions. Idem. p. 193-202 (1973).	R,δR,δR$_\perp$. 50 keV Ar; 100, 180 keV Kr $\rightarrow$ Si
541	Y. Akasaka, K. Horie: Channeling analysis and electrical behavior of boron implanted silicon. Idem. p. 145-57 (1973).	R,δR. 100 keV ^{11}B $\rightarrow$ Si
542	W. K. Hofker, H. W. Werner, D. D. Oosthoek, H. A. M. de Grefte: Experimental analysis of concentration profiles of boron implanted into silicon. Idem. p. 133-45 (1973).	R,δR. 30-70 keV B $\rightarrow$ Si
543	K. Wittmaack, J. Maul, F. Schulz: Energy dependence and annealing behaviour of boron range distributions in silicon. Idem. p. 119-31 (1973).	R,δR. 10-250 keV B $\rightarrow$ Si
544	M. Iwaki, K. Gamo, K. Masuda, S. Namba, S. Ishihara, I. Kimura: Concentration profiles of arsenic implantated in silicon. Idem. p. 111-18 (1973).	R,δR. 35-130 keV As $\rightarrow$ Si
545	J. C. Tsai, J. M. Morabito, R. K. Lewis: Arsenic implanted and implanted diffused profiles in silicon using secondary ion emission and differential resistance. Idem. p. 87-97 (1973).	R,δR. 40 keV As $\rightarrow$ Si
546	P. Blood, G. Dearnaley, M. A. Wilkens: The depth distribution of phosphorus ions implanted into silicon crystals. Idem. p. 75-85 (1973).	R,δR. 40-120 keV ^{32}P $\rightarrow$ Si (cryst. and amorph.)
547	M. W. Friedlander, D. Keefe, M. G. V. Menon: The range in G5 nuclear emulsion of protons with energies 87, 118 and 146 MeV. Nuovo Cimento 5, 461-72 (1957).	R. 87, 118, 146 MeV α $\rightarrow$ Emulsion
548	J. A. Borders: Helium ion stopping cross sections in bismuth, lead and tungsten. Rad. Effects 21, 165-69 (1974).	S. 0.4-1.9 MeV α $\rightarrow$ Bi, Pb, W
549	F. C. Gilber, H. H. Heckman, F. M. Smith: Ranges of 14 MeV protons in nuclear emulsion. Rev. Sci. Instr. 29, 404-05 (1958).	R. 14 MeV p $\rightarrow$ Emulsion
550	K. Ramavataram, D. I. Porat: Measurement of surface density of thin foils. Nucl. Instr. Methods 4, 239-42 (1959).	S. 3.72, 4.33 MeV α $\rightarrow$ Al, Ni, Ag, Au all rel. to air
551	P. G. Roll, F. E. Steigert: Characteristics of heavy ion tracks in nuclear emulsion. Nuclear Physics 16, 534-44 (1960).	R. 2-40 MeV He, 5-100 MeV ^{10}B, 5.5-110 MeV ^{11}B, 6-120 MeV ^{12}C, 7-140 MeV ^{14}N, 8-160 MeV ^{16}O, 8.5-190 MeV ^{19}F, 10-200 MeV ^{20}Ne $\rightarrow$ Emulsion

552	J. M. Fairfield, B. L. Crowder: Ion implantation doping of silicon for shallow junctions. Trans. Met. Soc. AIME *245*, 469-73 (1969).	R,δR. 70-280 keV B, P, 80-480 keV As $\rightarrow$ Si
553	R. B. J. Palmer, H. A. B. Simons: The experimental determination of the range-energy relations for alpha particles in water and water vapour and the stopping power of water and water vapour for alpha energies below 8.78 MeV. Proc. Phys. Soc. *74*, 585-98 (1959).	R. 1-8.78 MeV α $\rightarrow$ H_2O (gas. and liq.)
554	Yu. A. Vorob'ev: Range of nitrogen and beryllium ions in air. Zh. eksp. teor. Fiz. *35*, 1306-07 (1958). [Engl. Trans. Sov. Phys. JETP *8*, 912 (1959].	R. 4.9-9.5 MeV N, 3.1, 3.7 MeV Be $\rightarrow$ Air
555	H. Nakata: Energy loss of 1-10 MeV nitrogen ions in antimony. Phys. Rev. B *9*, 4654-59 (1974).	S. 1-10 MeV N $\rightarrow$ Sb
556	J. Catalá, J. Casanova: A photographic plate study of the uranium-235 radioactivity (in Spanish). An. Real. Soc. Espan. Fis. Quim. *56A*, 57-59 (1960).	R. 4.2, 4.4 MeV α $\rightarrow$ Emulsion
557	L. Winsberg, J. M. Alexander: Ranges and range straggling of Tb[149], At and Po. Phys. Rev. *121*, 518-28 (1961).	R,δR. 4-29 MeV [149]Tb, 4-15 MeV At, Po $\rightarrow$ Al, 4-9 MeV At, Po $\rightarrow$ Au
558	R. L. Hines: Ranges of 7.5 to 52 keV H^+_2, D^+_2, He^+, and Ne^+ ions in quartz. Phys. Rev. *120*, 1626-30 (1960).	R. 7.5-52 keV H^+_2, D^+_2, He^+, Ne^+ $\rightarrow$ SiO_2 (cryst.)
559	J. A. Davies, J. D. McIntyre, G. Sims: The range of Cs[137] ions of low keV energies in germanium. Can. J. Chem. *40*, 1605-10 (1962).	R. 4-20 keV [137]Cs $\rightarrow$ Ge.
560	G. Anianson: The integral stopping power of liquid hydrocarbons for 5.3 MeV α-particles. Kgl. Tekn. Högsk. Handl (Sweden). No. 178, 1-31 (1961).	R. 5.3 MeV α $\rightarrow$ 21 hydrocarbons
561	M. Ya. Gen, Yu. I. Petrov: The range in aluminium of 14.7 MeV protons. Atomnaya Energiya *7*, 473 (1959). [Engl. Trans. Reactor Sci. *13*, 91-92, (1960)].	R. 14.7 MeV p $\rightarrow$ Al
562	R. B. J. Palmer: The stopping power for alpha particles of ethyl alcohol and carbon tetrachloride in the liquid and vapour state. Proc. Phys. Soc. *78*, 766-73 (1961).	R. 1-8.9 MeV α $\rightarrow$ C_2H_5OH, CCl_4 (liq. and vap.)
563	V. A. J. van Lint, R. A. Schmitt, C. S. Suffredini: Range of 2 to 60 keV recoil atoms in Cu, Ag, and Au. Phys. Rev. *121*, 1457-63 (1961).	R. 2.4-57.5 keV Cu $\rightarrow$ Cu; 2.9-27.2 keV Ag $\rightarrow$ Ag; 6.1-15.1 keV Au $\rightarrow$ Au
564	Cao Xuan Chuan: Relations expérimentales parcours-énergie pour les ions légers de faible énergie dans les émulsions nucléaires. J. Phys. Radium *21*, 757-59 (1960).	R. 2.2-6.3 MeV [11]B, 2.2-5.0 MeV [13]C $\rightarrow$ Emulsion
565	G. Gérardin, R. Bilwes, D. Magnac-Valette: Pouvoir d'arrêt de l'or pour des particules alpha de 14.36 MeV. J. Phys. Radium *22*, 62-64 (1961).	R. 9.61-14.36 MeV α $\rightarrow$ Au

566	J. Blandin-Vial: Ralentissement des particules du thorium C et du thorium C' par des écrans d'or. C. R. Acad. Sci. *254*, 3842-44 (1962).	S. 6.05, 6.89 MeV $\alpha \rightarrow$ Au
567	U. Riezler, A. Rudloff: Ionisation und Energieverlust von Alpha-Teilchen in verschiedenen Gasen. Ann. Physik *18*, 224-45 (1955).	R. S rel. to air. 5.3 MeV $\alpha \rightarrow$ He, Ne, Ar, Kr, Xe, H_2, N_2, O_2, NH_3, CO, CO_2, NO, N_2O, CH_4, C_2H_6, C_3H_8, C_4H_{10}
568	W. Riezler, H. Schepers: Ionisation und Energieverlust von Alpha-Teilchen in verschiedenen Gasen. Ann. Physik *8*, 270-77 (1961).	R. S rel. to air 8.78 MeV $\alpha \rightarrow$ Air, He, Ne, Ar, Kr, H_2, N_2, O_2, CO, CO_2, CH_4, C_2H_6, C_3H_8, C_4H_{10}
569	J. P. Lonchamp: Étude par la méthode de la plaque photographique des ions Li accélérés. J. Phys. Radium *18*, 239-46 (1957).	R. 7.03-19.32 MeV p, 9.02-25.22 MeV $\alpha \rightarrow$ Emulsion
570	W. Primak, Y. Dayal, E. Edwards: Ion bombardment of silicon. J. Appl. Phys. *34*, 827-38 (1963).	R. 100 keV p $\rightarrow$ Si
571	A. Perez, P. Thevenard, C. H. S. Dupuy: Investigations on electronic stopping power in alkali halides by means of color center profiles, in S. Datz, B. R. Appleton, C. D. Moak (edts.): *Atomic Collisions in Solids*. Plenum N. Y. p. 47-56 (1975).	S. 28 MeV d, 56 MeV $\alpha \rightarrow$ LiF
572	C. D. Moak, B. R. Appleton, J. A. Biggerstaff, S. Datz, T. S. Noggle: Velocity dependence of the stopping power of channeled iodine ions. Idem. p. 57-62 (1975).	S. 21.6-31.25 MeV I $\rightarrow$ Ag (cryst.)
573	S. Datz, B. R. Appleton, J. A. Biggerstaff, M. D. Brown, H. F. Krause, C. D. Moak, T. S. Noggle: Charge state dependence of stopping power for oxygen ions channeled in silver. Idem. p. 63-73 (1975).	S,δS. 27.8-40.0 MeV O $\rightarrow$ Ag (cryst.) Charge state dep.
574	G. della Mea, A. V. Drigo, S. lo Russo, P. Mazzoldi, G. G. Bentini: Transmission energy loss of protons channeled in thin silicon single crystals of medium energy. Idem. p. 75-76 (1975).	S. 50-300 keV p $\rightarrow$ Si (cryst.) Chann. to random ratio
575	H. Schmidt-Böcking, G. Rühle, V. Bethge: A new method to determine the energy loss of heavy ions in solids. Idem. p. 77-83 (1975).	S. 7-38 MeV ^{16}O $\rightarrow$ Au; 9-30 MeV ^{32}S $\rightarrow$ Au; 8-30 MeV ^{16}O, ^{19}F $\rightarrow$ Ni
576	J. H. Barrett, B. R. Appleton, T. S. Noggle, C. D. Moak, J. A. Biggerstaff, S. Datz, R. Behrisch: Hyperchanneling. Idem. p. 645-48 (1975).	S,δS. 21.6-31.25 MeV I $\rightarrow$ Ag (cryst.)
577	G. Götz, K. D. Klinge, U. Finger: A. combination of dechanneling and energy measurements of protons in thin silicon single crystals. Idem. p. 693-716 (1975).	S. 0.7-1.8 MeV p $\rightarrow$ Si (cryst.) Chann. to random ratio.
578	J. P. Biersack, D. Fink: Channeling, blocking and range measurements using thermal neutron induced recutions. Idem. p. 737-47 (1975).	R,δR. 220 keV Li $\rightarrow$ Ag, Nb

579	F. H. Eisen, J. Bøttiger: Transmission energy spectra of channeled protons scattered in thin silicon films. Idem. p. 919-27 (1975).	S,δS. 200, 400 keV p → Si (cryst.)
580	S. Barkan: Differential energy loss measurements for α-rays in metal foils. Rev. Fac. Sci. Univ. Istanbul C *28*, 71-80 (1963).	S. 5-9 MeV α → Al, Ni, Au.
581	I. Hauser: Die Reichweite van 6.50 MeV-Protonen in der Agfa-K2-Emulsion. Exper. Tech. Phys. *11*, 126-29 (1963).	R. 6.5 MeV p → Emulsion
582	J. A. Neuendorfer, D. R. Inglis, S. S. Hanna: Angular yields of deuterons and alphas from the proton bombardment of beryllium. Phys. Rev. *82*, 75-80 (1951).	R. 200-900 keV p, 600-1500 keV d, 1200-2400 keV α, 750-1400 keV ^{6}Li → Emulsion
583	J. Csikai, P. Bornemisza, I. Hunyadi: Nuclear recoil in 14.8 MeV neutron reactions. Nucl. Instr. Methods, *24*, 227-28 (1963).	R. 1.95 MeV ^{27}Mg, 3.81 MeV ^{24}Na → Al
584	B. M. Nosenko, N. A. Strukov, M. D. Yagudaev: Luminescence of crystal phosphors on excitation with ions. Optika i Spektroshopika *3*, 351-55 (1957) (in Russian).	R. 2.4-6 keV Li, Na, K, Cs → ZnS
585	W. H. Barkas, P. H. Barrett, P. Cüer, H. H. Heckman, F. M. Smith, H. K. Ticho: The range-energy relation in emulsion. I. Range measurements. Nuovo Cimento *8*, 185-200 (1958).	R. 2.5, 5 MeV t, 5, 5.5 MeV d, 21.2 MeV p, 5 MeV α, 5, 10 MeV ^{3}He, 36.6 MeV μ^+, 200-700 MeV π^+ → Emulsion
586	J. P. Lonchamp, G. Robin: Sur la relation parcours-énergie des ions ^{8}Li dans les émulsions nucléaires. C. R. Acad. Sci. *246*, 748-50 (1958).	R. 20.6-22.4 MeV ^{8}Li → Emulsion
587	Nguyen-Huu-Tri: Sur la relation parcours-énergie des ions $^6_3$Li dans les émulsions nucléaires. C. R. Acad. Sci. *250*, 2016-18 (1960).	R. 0.75-2.0 MeV ^{6}Li → Emulsion
588	J. P. Lonchamp: Sur la relation parcours-énergie des ions ^{7}Li dans les émulsions nucléaires Ilford C$_2$. C. R. Acad. Sci. *244*, 1486-88 (1957).	R. 9.02-25.22 MeV ^{7}Li → Emulsion
589	F. Demichelis: α-particles straggling in mica and aluminum. Nuovo Cimento *13*, 562-71 (1959).	δS. 5.3 MeV α → Al, mica
590	E. S. Anashkina: Range-energy relation of protons in NiKFi-Ya2 emulsion. Pribory Tekh. Eksper. No. 4, 148 (1961). [Engl. trans. Instr. Exp. Tech. No. 4, 772-73, (1961)].	R. 2.47, 14.2 MeV p → Emulsion
591	I. Hauser: Dependence of the proton range on energy in NiKFi-Ya2 emulsion. Pribory Tekh. Eksper. No. 6, 60-64 (1963). [Engl. trans. Instr. Exp. Tech. No. 6, 1172-73, (1963)].	R. 6.47 MeV p → Emulsion
592	C. M. Portner, R. B. Moore: A precise measurement of the range of 100-MeV protons in aluminum. Can. J. Phys. *43*, 1904-14 (1965).	R. 100 MeV p → Al

593	S. Datz, T. S. Noggle, C. D. Moak: Channeling effects on the energy loss of high energy (20-80 MeV) ^{79}Br and ^{127}I ions in gold. Nucl. Instr. Methods *38*, 221-30 (1965).	S,δS. 20-80 MeV ^{79}Br, ^{127}I → Au (cryst.)
594	Y. Y. Chu, L. Friedman: Use of the d-d-reactions to investigate penetration of 20 keV denteron in gold and aluminum. Nucl. Instr. Methods *38*, 254-59 (1965).	R,δR. 20 keV/nucleon D$^+$, D$^+_2$, D$^+_3$ → Al, Au
595	H. E. Wegner, B. R. Appleton, C. Erginsoy, W. M. Gibson: Axial and planar effects in the energy loss of protons in silicon single crystals. Phys. Letters *19*, 185-86 (1965).	S,δS. 4.85 MeV p → Si (cryst.)
596	F. Jähnig, J. Kalus: Messung de anisotropen Abbremsung von Chromatomen in Energiebereich 20-90 eV in einem Vanadiumeinkristall mit der Kernfluoreszensmethode. Z. Naturforschg. *20a*, 387-90 (1965).	S. 20-90 eV Cr → V (cryst.)
597	T. R. Ophel, J. M. Morris: Measurement of the energy distribution of charged particles after passage through a thin foil. Phys. Letters *19*, 245-47 (1965).	δS. 1.0l MeV p → C
598	O. Fiedler, D. Ulrich: Das relative Bremsvermögen einiger Substanzen für α-Teilchen bis 5 MeV. Z. Physik *200*, 493-98 (1967).	S. 0.3-5 MeV α → Al, Ag, Au, Zapon, Paraffine.
599	E. T. Shipatov, B. A. Kononov: Investigation of the channeling of protons in single crystals of ionic compounds and semiconductors. Izv. vuz Fiz. No. 9, 52-56 (1968). [Engl. trans. Soviet Phys. J. No. 9, 46-49, (1968)].	S,δS. 4.7, 6.72 MeV p → NaCl, KCl, KBr, Si, Ge (all cryst.)
600	L. T. Chadderton, M. G. Anderson: Energy structure in the axial channeling of 30 keV protons through gold. Phys. Letters *27A*, 665-66 (1968).	S, δS. 30 keV p → Au (cryst.)
601	A. R. Sattler, F. L. Vook: Channeling in zinc-blende lattices: energy-loss studies for hydrogen and helium ions in InAs, GaSb, AlSb, and InSb. Phys. Rev. *175*, 526-32 (1968).	S. 4.5-6.9 MeV p → InAs, 1.9-8.0 MeV p, 2.5-8.0 MeV d, 5.4-13.9 MeV ^{3}He → GaSb, 3.4-6.7 MeV p → InSb, 3.9 MeV p → AlSb (all (cryst.)
602	J. Rémillieux, J. J. Samueli, A. Sarazin: Étude des effets directionnels dans la transmission de protons de 2 MeV a travers un monocristal de silicium. J. Physique *28*, 832-38 (1967).	S, δS. 2 MeV p → Si (cryst.)
603	K. E. Manchester: Radiotracer study of ion implanted profile build-up in silicon substrates. J. Electrochem. Soc. *115*, 656-59 (1968).	R, δR. 40 keV ^{31}P, ^{32}P → Si (cryst.)
604	E. T. Shipatov, V. A. Kononov, V. P. Ivakin: Orientation dependence of energy loss of fast protons in a KBr single crystal. Izv. VUZ. Fiz. No. 2, 136-38 (1968). [Engl. trans. Soviet Phys. J. No. 2, 91 (1968).]	S, δS. 6.72 MeV p → KBr (cryst.)

605	W. Neuwirth, U. Hauser, E. Kuhn: Energy loss of charged particles in matter. I. Experimental method and velocity dependence of the energy loss of lithium ions. Z. Physik *220*, 241-64 (1969).	S. 100-800 keV Li → B_4C, B, H_2O, H_3BO_3, MoB, WB
606	M. W. Thompson, G. W. Neilson: Effects of inner shell excitations on the stopping power of solids for heavy ions. Phys. Letters *49A*, 151-53 (1974).	R. 100 keV Ba, La, Ce, Pr, Nd, Sm → Al
607	R. J. Blewer: Proton backscattering as a technique for light ion surface interaction studies in CTR materials investigations. J. Nucl. Mater. *53*, 268-75 (1974).	R, δR. 50-150 keV He → Cu, 50 keV He → Ti, V, Nb
608	J. P. Biersack, D. Fink: Damage and range profiles of lithium implanted into niobium. J. Nucl. Mater. *53*, 328-31 (1974).	R. 220 keV ^{7}Li → Nb
609	R. Gähler: Zerstäubung von Gold mit 14 MeV Neutronen. Diplomarbeit, Technische Universität München (1974).	R. 71 keV Au → Au, 143 keV Nb → Nb
610	H. Bach. I, Kitzmann, H. Schröder: Sputtering yields and specific energy losses of Ar^+ ions with energies from 5 to 30 keV at SiO_2. Rad. Effects *21*, 31-36 (1974).	S. 5-30 keV Ar → SiO_2
611	J. S. Williams, W. A. Grant: High resolution Rutherford backscattering and its application to ion range and ion collection measurements. Rad. Effects *25*, 55-56 (1975).	R, δR. 20-80 keV Kr, Xe, Cs, Dy, Au, Pb, Bi → Si, Al
612	G. Sidenius: Systematic stopping cross section measurements with low energy ions in gases. Kgl. Danske Videnskab. Selskab. Mat. Fys. Medd. *39*, No. 4, 1-32 (1974).	S. 0.6-70 keV H, He, 2-120 keV ^{6}Li, ^{7}Li, 3-120 keV Be, B, C, N, O, F, Ne → CH_4
613	F. Brown, G. C. Ball, D. A. Channing, L. M. Howe, J. P. S. Pringle, J. L. Whitton: Ranges of heavy ions. Nucl. Instr. Methods *38*, 249-53 (1965).	R, δR. 20-150 keV Xe → Au, W; 20-80 keV Au → Au; 20-80 keV Xe → Si; 20-150 Kr → Al; 23 keV Na, 40 keV K, 80 keV Kr, Rb, 125 keV Xe, Cs, 185 keV Hg → Al, Si, W, Au (all targets cryst.); 0.5-100 keV ^{125}Xe → Ta_2O_5 40 keV ^{133}Xe → UO_2
614	W. K. Hofker, H. Werner, D. P. Oosthoek, N. J. Koeman: Boron implantations in silicon: A comparison of charge carrier and boron concentration profiles. Appl. Phys. *4*, 125-33 (1974).	R, δR. 70 keV B → Si
615	H. Müller, H. Kranz, H. Ryssel, K. Schmid: Electrical and backscattering measurements of arsenic implanted silicon. Appl. Phys, *4*, 115-23 (1974).	R. 150-200 keV As → Si (cryst.)
616	R. P. Gittins, D. V. Morgan, G. Dearnaley: The application of the ion microprobe analyzer for the measurements of the distribution of boron ions implanted into silicon crystals. J. Phys. D: Appl. Phys. *5*, 1654-63 (1972).	R, δR. 200 keV Na, 110-400 keV B → Si (cryst. chann. and random)

617	J. A. Cairns, R. S. Nelson, J. S. Briggs: The use of the ion-induced x-rays to investigate the concentration distribution and atom location of boron-implanted silicon, in I. Ruge and J. Graul (edts.): *Ion Implantation in Semiconductors*. Springer, Berlin, p. 299-306 (1970).	R. 100-300 keV B → Si
618	J. Consigny: Pouvoir d'arret de quelques métaux pour les rayons alpha. C. R. Acad. Sci. *183*, 127-29 (1926).	S rel. to air. 5.3 MeV α → Al, Cu, Ag, Au
619	C. L. Shepard, L. E. Porter: Stopping power of several composite materials for 2.5 and 3.5 MeV deuterons and 5.5 MeV α particles. Phys. Rev. B*12*, 1649-57 (1975).	S. 2.4, 3.5 MeV d → Havar, μ-metal, permalloy. 5.4 MeV α → havar, μ-metal, mylar, teflon
620	I. G. Guerdtsiteli, A. I. Guldamashvili, E. M. Disamidze, S. A. Zaslavsky, A. N. Kalinin, M. A. Kumakhov, V. A. Muralev: Implantation angle influence on penetration of boron channeled ions into silicon. Rad. Effects *19*, 171-74 (1973).	R, δR. 30-200 keV B → Si (cryst.)
621	K. L. Dunning, G. K. Hubsher, J. Comas, W. H. Lucks, H. L. Hughes: Depth profiles of aluminum and sodium near surfaces: Nuclear resonance method. Thin Solid Films *19*, 145-56 (1973).	R, δR. 60 keV Al → SiC, 20, 60 keV Na → SiO_2
622	J. Tripier, G. Remy, J. Rarlarosy, M. Debauvais, R. Stein, D. Huss: Range and energy loss rate for heavy ions in makrofol and cellulose nitrate. Nucl. Instr. Methods *115*, 29-46 (1974).	R. Fiss. fragm. → makrofol, cellulose nitrate
623	S. A. Gabriele, P. Giusti, T. Massami, F. Palmonari, G. Valenti, A. Zichichi: Observation of relativistic rise in the energy loss in plastic scintillator. Nucl. Instr. Methods *113*, 465-68 (1973).	S. 13-20 GeV/c p → plast. scint.
624	R. Loew: Charge-state and energy distributions of xenon ions in carbon. Nucl. Instr. Methods *118*, 504-08 (1974).	S. 300-700 keV Xe → C
625	D. Jeanne, P. Lazeyras, I. Lehraus, R. Mathewson, W. Tejessy, M. Adlerholz: High energy particle identification using multilayer proportional counters. Nucl. Instr. Methods *111*, 287-300 (1973).	S. 1.5-16 GeV/c p, π, K → Ar + 5% CH_4
626	M. Adlerholz, P. Lazeyras, I. Lehraus, R. Mathewson, W. Tejessy: High-resolution ionization measurements in the region of relativistic rise. Nucl. Instr. Methods *118*, 419-30 (1974).	S, δS. 9 GeV/c p, π → Ar + 5% CH_4, 60% He + 30% Ar + 10% CH_4, Kr + 5% CH_4
627	H. Schmidt-Böcking, G. Rühle, K. Bethge: The determination of the differential energy loss of heavy ions backscattered from an infinitely thick solid target. Nucl. Instr. Methods *118*, 357-60 (1974).	S. 7-35 MeV ^{16}O → Ni, Au
628	V. V. Avdeichikov, E. A. Ganza, O. V. Lozhkin: Energy-loss fluctuation of heavy charged particles in silicon absorbers. Nucl. Instr. Methods *118*, 247-52 (1974).	δS. 7.7 MeV α, 255 MeV Ar → Si

629	P. F. Engel, J. A. Borders, F. Chernow: Stopping power of cadmium sulfide for helium ions. J. Appl. Phys. *45*, 38-42 (1974).	S. 0.96-5.3 MeV $\alpha \rightarrow$ CdS
630	J. Whitton: The dependence of electronic stopping cross section of ^{42}K on different target materials. Can. J. Phys. *52*, 12-16 (1974).	R_{max}. 55 keV ^{42}K $\rightarrow$ Cu, Ag, Au, V, Mo, Nb, Ta, W (all cryst.)
631	C. D. Moak, S. Datz, B. R. Appleton, J. A. Biggerstaff, M. D. Brown, H. F. Krause, T. S. Noggle: Influence of ionic charge state on the stopping power of 27.8 and 40 MeV oxygen ions in the [011] channel of silver. Phys. Rev. B*10*, 2681-86 (1974).	S. Fixed charge state. 27.8, 40.0 MeV O $\rightarrow$ Ag (cryst.)
632	A. Bontemps, E. Ligeon, J. Fonteville: Energy loss and projected range of alpha particles in zinc telluride. Rad. Effects *21*, 181-84 (1974).	S. 0.3-2.0 MeV $\alpha \rightarrow$ ZnTe
633	A. Rudnev, K. S. Shyskin, E. I. Sirotinin, A. F. Tulinov: The determination of energy losses of channelled particles from the backscattering spectra. Rad. Effects *22*, 29-33 (1974).	S. 6.3 MeV p $\rightarrow$ W (cryst.)
634	E. Leminen, A. Fontell: Stopping power of Ti, Mo, Ag, Ta and W for 0.5-1.75 MeV ^{4}He ions. Rad. Effects *22*, 39-44 (1975).	S. 0.5-1.75 MeV $\alpha \rightarrow$ Ti, Mo, Ag, Ta, W
635	T. A. Lagerlund, B. Blecher, K. Gotow, R. Keller, W. C. Lam: Range and multiple-scattering measurements of low-energy muons. Nucl. Instr. Methods *120*, 521-24 (1974).	R. 9-25 MeV $\mu^+ \rightarrow$ Al, steel, polyethylene
636	G. Beachemin, R. Drouin: Angular behaviour of stopping powers in carbon for heavy ions below 250 keV; in J. L. Duggan, I. L. Morgan (edts.): *Proc. 3 rd Int. Conf. Appl. of Small Accelerators*. ERDA Technical Information Center. CONF 741040-P1, p. 336-46 (1975).	S. Dep. on scatt. angle 50-250 keV Ar $\rightarrow$ C
637	A. Nomura, S. Kiyono, M. Kamayama: Penetration of low energy helium ions through copper thin films. Jap. J. Appl. Phys. *13*, 1159-60 (1974).	S. 6-16 keV He $\rightarrow$ Cu
638	H. Okabayashi, D. Shinoda: Range and standard deviation of ion-implanted phosphorus in silicon. Jap. J. Appl. Phys. *13*, 1187-88 (1974).	R. 50-145 keV ^{31}P $\rightarrow$ Si
639	S. M. Myers, S.T. Picraux, T. S. Prevender: Study of Cu diffusion in Be using ion backscattering. Phys. Rev. B *9*, 3953-64 (1974).	R. 100 keV Cu $\rightarrow$ Be
640	T. W. Sigmon, W. K. Chu, H. Müller, J. W. Mayer: Analysis of arsenic range distributions in silicon. Applied Physics *5*, 347-50 (1975).	R, δR. 50-250 keV As $\rightarrow$ Si
641	E. M. Zarutskii: Penetration of heavy ions into copper, silver and gold. Fiz. Tverd. Tela *11*, 1684-89 (1969). [Engl. trans. Sov. Phys. Solid State *11*, 1362-65 (1969)].	S. 9-20 keV Rb, 11-19 keV Cs $\rightarrow$ Cu; 12-17 keV Rb, 16-20 keV Cs $\rightarrow$ Ag, Au

642	S. Hughes: The range of 5-50 keV heavy ions in various gases. Phys. Med. Biol. *12*, 565-71 (1967).	R. 5-50 keV H^+ → Ar, CO_2, N_2, CH_4, C_2H_5, C_2H_4, C_3H_8, C_4H_{10}. 5-30 keV N^+ → CH_4
643	D. Schalch, A. Scharmann: Eindringtiefen von Ionen in CaF_2-und Rb-Aufdampfschichten. Z. angew. Phys *29*, 111-13 (1970).	R. 10-80 keV H, He, Ne, Ar, Kr, Xe → CaF_2, RbJ
644	B. L. Crowder: The role of damage in the annealing characteristics of ion implanted Si. J. Electrochem. Soc. *117*, 671-74 (1970).	R, δR. 280 keV ^{31}P → Si
645	P. J. McNulty, F. J. Congel: Restricted energy loss by extremely relativistic particles. Phys. Rev. D *1*, 3041-44 (1970).	S. rel. to min. 5 GeV π, 5-24 GeV p → Emulsion
646	B. A. Kononov, V. K. Struts: Channeling of protons in silicon at different temperatures. Izv. Vuz. Fiz. No. 6, 60-63 (1970). [Engl. trans. Sov. Phys. J. *13*, 738-61 (1970)].	S, δS. 6.72 MeV p → Si (cryst.)
647	G. A. Zhukova, V. S. Kesselman, V. N. Mordkovich, G. F. Zabotina: The slowing down of low energy protons in SiO_2 films. Zh. Eksp. Teor. Fiz. *59*, 414-18 (1970). [Engl. trans. Sov. Phys. JETP *32*, 226-28 (1971).]	R. 15-50 keV p → SiO_2
648	O. Meyer, J. W. Mayer: Analysis of Rb and Cs implantations in silicon by channeling and Hall effect measurements. Solid-State Electronics *13*, 1357-62 (1970).	R, δR. 5-50 keV Rb, Cs → Si
649	L. N. Large, H. Hill, M. P. Ball: Profiles of high conductivity shallow layers in silicon produced by boron ion implantations. Int. J. Electronics *22*, 153-64 (1967).	R, δR. 25-125 keV B → Si
650	S. Roosild, R. Dolan, B. Buchanan: Semiconductor doping by high energy 1-2.5 MeV ion implantation. J. Electrochem. Soc. *115*, 307-11 (1968).	R, δR. 1-2.5 MeV B, 1-1.6 MeV N, 1 MeV p → Si
651	Yu. V. Gott, V. G. Tel'kovsky: Deceleration of slow hydrogen and deuterium ions in a thin silver foil. Fiz. Tverd. Tela *9*, 2221-24 (1967). [Engl. trans. Sov. Phys. Solid State *9*, 1741-44 (1968).]	S. 0.2-40 keV p,d → Ag
652	G. Sidenius: Measurement of dE/dR in gases with low energy heavy particles; in M.R.C. McDowell (edt.) *Atomic Collision Processes*, North-Holland, Amsterdam, p.709-16 (1964).	S. 20-50 keV ^{35}Cl, ^{69}Ga, 30-50 keV ^{90}Zr, ^{121}Sb, ^{208}Pb, 40,50 keV ^{56}Fe, 50 keV ^{40}Ca, ^{74}Ge, ^{238}U → H_2
653	É. T. Shipatov, B. A. Kononov: Energy distribution of 6.72 MeV protons passing through monocrystals. Atomnaya Energiya *25*, 439-40 (1968). [Engl. trans. Sov. Atom. Energy *25*, 1254-55 (1968).]	S, δS. 6.72 MeV p → NaCl, KCl, KBr, Si, Ge (all cryst.)
654	É. T. Shipatov: Energy and angular distributions of protons transmitted by germanium and silicon single crystals along (110) and (100) channels in the crystal lattice. Fiz. Tekh. Poluprovodnikov *2*, 1690-91 (1968). [Engl. trans. Sov. Phys. Semicond. *2*, 1408-09 (1969).]	S, δS. 6.72 MeV p → Si, Ge (both cryst.)

655	D. Apel, U. Müller-Jahreis, S. Schwabe: On the Z_2-dependence of electronic stopping cross section. Phys. Stat. Sol. (a) *3*, K173-75 (1970).	S. 10-100 keV Li → Si, V, Cr, Fe, Ge, Se
656	E. I. Zorin, P. V. Pavlov, D. I. Tetelbaum: Investigation of boron atom distribution in silicon doped by ion bombardment method. Fiz. Tverd. Tela *9*, 3642-44 (1967). [Engl. trans. Sov. Phys. Solid State *9*, 2874-76 (1968).]	R, δR. 20-60 keV B → Si
657	C. C. Hanke, H. Bichsel: Precision energy loss measurements for natural alpha particles in argon. Kgl. Danske Videnskab. Selskab. Mat. Fys. Medd. *38*, No. 3, 1-29 (1970).	S. 1-9 MeV α → Ar
658	W. Bernstein, A. J. Cole, R. L. Wax: Penetration of 1-20 keV ions through thin carbon foils. Nucl. Instr. Methods *90*, 325-28 (1970).	S. 1-20 keV H, O, He, Li, N, Ne, K → C
659	R. S. Nelson, J. A. Cairns: Antimony implanted silicon: a comparison between the total implanted concentration profile and the donor concentration profile. Rad. Effects *6*, 131-34 (1970).	R, δR. 100 keV Sb → Si
660	T. Jokic: Ranges of 20-80 keV xenon ions in copper crystals; in *Proc. 9th Int. Conf. Phenom. Ionized Gases*, Bucharest, p. 96 (1969).	R, δR. 20-80 keV Xe → Cu (cryst.)
661	H. Dannheim, B. Gaissmair, D. Kroniger, K. H. Katchera, B. Seidler: Determination of dE/dx for 3-17 MeV alpha particles from energy loss measurements in lanthanum magnesium nitrate single crystals. Z. Physik *241*, 130-37 (1971).	S. 3-17 MeV α → $La_2Mg_3(NO_3)_{12} \cdot 24H_2O$
662	R. Skoog, K. Augenlicht-Jakobson: Elastic and electronic stopping cross sections for sodium and argon projectiles in carbon. Rad. Effects *27*, 143-149 (1976).	S. Dep. on scatt. angle. 50-150 keV Na, 50-300 keV Ar → C
663	P. Hvelplund: Energy loss and straggling of 100-500 keV $_{90}$Th, $_{82}$Pb, $_{80}$Hg, and $_{64}$Gd in H_2. Phys. Rev. A*11*, 1921-27 (1975).	S, δS. 100-500 keV Gd, Hg, Pb, Th → H_2
664	K. W. Jones, H. W. Kraner: Energy lost to ionization by 254-eV ^{73}Ge atoms stopping in Ge. Phys. Rev. A*11*, 1347-53 (1975).	S. $\eta(\varepsilon)$ 254 eV ^{73}Ge → Ge
665	D. A. Sykes, S. J. Harris: The anomalous straggling of alpha particles in aluminum compared to other absorbers. Nucl. Instr. Methods *101*, 423-25 (1972).	δS. 6.8 MeV α → Al, Cu, Ag, Au, Melinex
666	J. S. Huebner, L. L. Skolil: The residual energy and stopping power of ^{210}Po alpha particles in air, CO_2, and He. Am. J. Phys. *40*, 1177-78 (1972).	S. 3-5.3 MeV α → Air, CO_2, He
667	N. A. Baily, J. E. Steigerwalt, J. W. Hilbert: Frequency distributions of energy loss by fast charged particles after passage through large thicknesses of tissue-equivalent materials. Health Physics *22*, 497-502 (1972).	δS. 600 MeV p → Plastics

668	Yu. V. Bulgakov, V. S. Nikolaev, V. I. Shulga: The experimental determination of the impact parameter dependence of inelastic energy loss of channeled ions. Phys. Letters *46A*, 477-78 (1974).	S, δS. 1.15, 1.75 MeV p, 5.7 MeV N $\rightarrow$ Si (cryst.)
669	H. Lutz, R. Ambros, C. Mayer-Böricke, K. Reichelt, M. Rogge: Experimental evidence of fine structure in channeling lines. Z. Naturforschg. *26a*, 1105-08 (1971).	S, δS. 2 MeV α $\rightarrow$ Au (cryst.)
670	W. Brandt, A. Ratkowski, R. H. Ritchie: Energy loss of swift proton clusters in solids. Phys. Rev. Letters *33*, 1325-28 (1974).	S rel. to H$^+$. 60-300 keV H$^+$, 75, 150 keV H$_2^+$, 60-100 keV H$_3^+$ $\rightarrow$ C, Au
671	V. M. Gusev, N. P. Busharov, K. D. Demakov, Yu. G. Kozlow: Effects of channeling on the distribution of electrically active boron and phosphorus atoms in silicon single crystals. Dokl. Akad. Nauk SSSR *19*, 319-22 (1971). [Engl. trans. Sov. Phys. Doklady *19*, 213-15 (1971).]	R, δR. 100 keV B, 150 keV P $\rightarrow$ Si (cryst. and random)
672	K. W. Jones, H. W. Kraner: Stopping of 1- to 1.8 keV ^{73}Ge atoms in germanium. Phys. Rev. C *4*, 125-29 (1971).	S. $\eta(\varepsilon)$ 1-1.8 keV ^{73}Ge $\rightarrow$ Ge
673	B. R. Appleton, S. Datz, C. D. Moak, M. T. Robinson: Energy-loss spectra of channeled iodine and oxygen atoms in gold. Phys. Rev. B*4*, 1452-57 (1971).	S, δS. 15, 21.6 MeV ^{127}I, 10 MeV ^{16}O $\rightarrow$ Au (cryst.)
674	D. A. Sykes, S. J. Harris: Energy straggling of alpha particles in thick absorbers. Nucl. Instr. Methods *94*, 39-44 (1971).	δS. 5.5 MeV α $\rightarrow$ Al
675	W. K. Chu, D. Powers: Energy loss of α particles in noble gases from 0.3 to 2.0 MeV. Phys. Rev. B*4*, 10-15 (1971).	S. 0.3-2.0 MeV α $\rightarrow$ He, Ne, Ar, Kr, Xe
676	H. Rasekhi, F. A. White: A transmission method for measuring the stopping power of low energy ions in solids. *10th Natl. Meeting Soc. Appl. Spectroscopy*. The Society for Applied Spectroscopy. New York, p. 70 (1971).	R. 5-25 keV Li, Na $\rightarrow$ Ni
677	A. A. Bednyakov, V. G. Ignatov, A. F. Tulinov, Yu. N. Shustikov: Multiple scattering and energy losses of helium ions in aluminum for energies below 300 keV (in Russian). Vestn. Mosk. Univ. Fiz. Astron. No. 4, 402-10 (1971).	S. 300 keV α $\rightarrow$ Al
678	M. R. Arora, R. Kelly: A radiochemical technique for determining depth distributions in Mo. J. Electrochem. Soc. *119*, 270-74 (1972).	R, δR. 10 keV Kr $\rightarrow$ Mo
679	R. D. Edge, W. R. Hedrick, R. L. Dixon: A comparison of proton channeling in the <111> direction for BaF$_2$ and CaF$_2$. Rad. Effects *12*, 97-103 (1972).	S, δS. 300-400 keV p $\rightarrow$ BaF$_2$, CaF$_2$ (both cryst.)
680	F. Dolle, J. Schultz, P. Chevallier, G. Sutter: Étude de ralentissement des particules α de 6.620 MeV dans des absorbants gazeux (Ne, Ar, Kr) dans les domaine des faibles pertes d'énergie. J. Physique *32*, 397-403 (1971).	S, δS. 6.62 MeV α $\rightarrow$ Ne, Kr, Ar
681	J. L. Whitton, G. R. Bellavance: Ion implantation of sulphur into GaAs, GaP and Ge monocrystals. Rad. Effects *9*, 127-31 (1971).	R, δR. 20-40 keV S $\rightarrow$ GaAs (cryst. axial and random)

682	V. V. Makarov, N. N. Petrov: Investigation of the slowing down of positive ions in silicon carbide. Fiz. Tekh. Polupro-pov. *5*, 510-13 (1971). [Engl. trans. Sov. Phys. Semicond. *5*, 447-49 (1971).]	R. $\eta(\varepsilon)$. 1-20 keV H, Li, 2-20 keV D, He, Na, 3-20 keV K $\rightarrow$ SiC
683	E. M. Zarutskii: Retardation of lithium ions in silver. Fiz. Tverd. Tela *13*, 629-30 (1971). [Engl. trans. Sov. Phys. Solid State *13*, 516-17 (1971).]	S. 5-16 keV Li $\rightarrow$ Ag
684	J. S. Williams: Range and stopping power effects obtained from high resolution Rutherford backscattering analysis of implanted targets. Rad. Effects *22*, 211-13 (1974).	R, δR. 40 keV Pb $\rightarrow$ Si
685	R. Bimbot, D. Gardes: Mesures des parcours de recul de noyaux émetteurs alpha produits par réactions nucléaires d'apres le déplacement en énergie du spectre. Nucl. Instr. Methods *109*, 333-40 (1973).	R. 3.2-4.2 MeV ^{211}At $\rightarrow$ Al. S, δS. 7.45 MeV $\alpha \rightarrow$ Al
686	D. Phillips, J. P. S. Pringle: Surface effects in the measure-ment of range profiles by oxide dissolution. J. Electrochem. Soc. *120*, 1067-66 (1973).	R, δR. 5, 40 keV Tl, Au $\rightarrow$ Ta_2O_5
687	N. N. Demidovich, I. E. Nakhutin, V. G. Shatunov: Energy-loss and range straggling of α particles from Pu^{238} in some gases. Yad. Fiz. (SSSR) *18*, 133-41 (1973). [Engl. trans. Sov. J. Nucl. Phys. *18*, 70-73 (1974).]	S, δS. 3-5.5 MeV $\alpha \rightarrow CH_4$, CO_2, CF_4, CF_2Cl_2, SF_6
688	R. Kulessa, P. H. Barker, P. M. Cockburn, H. P. Seiler, P. Marmier: Determination of the range of ^{13}C and ^{19}F ions in Ni and Ta. Helv. Phys. Acta *46*, 52 (1973).	R. 2-12 MeV ^{13}C, ^{19}F $\rightarrow$ Ni, Ta
689	K. Kubo: Effects in the proton bombardment of NaF crys-tals. I. Measurement of the average projected range. J. Phys. Soc. Jap. *33*, 1401-06 (1972).	R. 0.6-1.0 MeV p $\rightarrow$ NaF
690	H. Nishi, T. Sakurai, T. Furuya: Investigation of boron im-planted silicon by backscattering method. Defect and impur-ity distribution. Fujitsu Sci. and Tech. J. *8*, 123-35 (1972).	R, δR. 100 keV B $\rightarrow$ Si
691	P. J. Walsh, F. Pendergrass: Energy loss of alpha particles in tissue equivalent plastic. Health Physics *23*, 701-04 (1972).	S. 1-5.5 MeV $\alpha \rightarrow$ TEF (Plastic)
692	Y. Akasaka, K. Horie, K. Yoneda, T. Sakurai, H. Nishi, S. Kawabe, A. Tohi: Depth distribution of defects and impuri-ties in 100-keV B^+ ion implanted silicon. J. Appl. Phys. *44*, 220-24 (1973).	R, δR. 100 keV B $\rightarrow$ Si
693	K. Sone, F. Fukusawa: Transmission of fast protons through Si single crystals. Mem Fac. Eng. Kyoto Univ. *34*, 325-32 (1972).	S, δS. 3 MeV p $\rightarrow$ Si (cryst.)
694	R. B. J. Palmer: The stopping power of organic liquids for alpha particles over the energy range 1-8 MeV. J. Phys. B: Atom. and Molec. Phys. *6*, 384-92 (1973).	S. 1-8 MeV $\alpha \rightarrow CCl_4$, Liq. and sol. hydrocarbons

695	J. A. Cairns, D. F. Holloway, R. S. Nelson: Measurement of implanted boron concentration profiles in silicon by the use of heavy ion x-ray excitation; in *Proceedings of the European Conference on Ion Implantation*. Peter Peregrinus, Publisher, Stevenhage, England, p. 203-06 (1970).	R, δR. 40, 100 keV B $\rightarrow$ Si
696	M. A. Wilkins, G. Dearnaley: The distribution of P[32] channelled into indium antimomide. Idem, p. 193-97 (1970).	R, δR. 40 keV [32]P $\rightarrow$ InSb (cryst., axial and rand.)
697	G. Eldridge, P. K. Govind, D. A. Nieman, F. Chernow: Radiation damage studies of bismuth ion-implanted CdS. Idem. p. 143-48 (1970).	R, δR. 25 keV Bi $\rightarrow$ CdS
698	D. P. Lecrosnier, G. P. Pelous: High energy boron implantation into silicon. Idem, p. 106-06 (1970).	R, δR. 1-2.5 MeV [11]B $\rightarrow$ Si
699	A. Izmen, O. Birgal, N. K. Aras: Ranges of [99]Mo and [140]Ba in several stopping media from the spontaneous fission of [252]Cf. J. Inorg. and Nucl. Chem. *36*, 25-29 (1974).	R. 179.4 MeV [99]Mo 188 MeV [140]Ba $\rightarrow$ Al, Ni, Cu, Pd
700	V. P. Zrelov, S. P. Kruglov, K. F. Mus, V. D. Savelyev, P. Shulek: Determination of the average excitation I_{Cu} of the copper ion from the ionization range of 660 MeV protons. Yad. Fiz. (SSSR) *19*, 1276-81 (1974). [Engl. trans. Sov. Journ. Nucl. Phys., *19*, 653-55 (1974)].	R. 660 MeV p $\rightarrow$ Cu
701	V. N. Lepeshinskaya, E. M. Zarutskii, A. D. Artemov: Diffusion method of determining ion penetration depth in refractory metals. Radiotekh. i. Elektronika *18*, 886-88 (1973). [Engl. trans. Rad. Eng. and Electron. Phys. *18*, 649-51, (1973).]	R. 4-9 keV Cs $\rightarrow$ W
702	A. Golanski, R. Dybczynski: Implantation of ions. II. Study on the distributions of effective ranges of phosphorus atoms implanted into silicon monocrystal (in Polish). Nukleonika *18*, 351-66 (1973).	R, δR. 16-60 keV P $\rightarrow$ Si
703	H. Blok, F. M. Kiely, B. D. Pate, F. Hanappe, J. Pelier: Further measurement of the track length of heavy ions in mica. Nucl. Instr. Methods *119*, 307-12 (1974).	R. 2.7-40 MeV Al, 4-60 MeV Ar, Ca, 5-80 MeV Cr, 6-85 MeV Ni, 8-120 MeV Se, Kr, 11-160 MeV Ag $\rightarrow$ Mica
704	J. M. Harris, M-A. Nicolet: Energy straggling of [4]He ions below 2.0 MeV in Al, Ni and Au. Phys. Rev. B*11*, 1013-19 (1975).	δS. 1-2 MeV α $\rightarrow$ Al, Ni, Au
705	F. Cembali, R. Galloni, F. Mousty, R. Rosa, F. Zignani: Doping and radiation damage profiles in silicon along the [110] axis. Rad. Effects *21*, 255-64 (1974).	R, δR. 200 keV P $\rightarrow$ Si (cryst.)
706	P. Blood, G. Dearnaley, M. Wilkens: The depth distribution of phosphorus ions implanted into silicon crystals. Rad. Effects *21*, 245-51 (1974).	R, δR. 40-120 keV [32]P $\rightarrow$ Si (cryst. and amorph.)
707	G. W. Neilson, B. W. Farmery, M. W. Thompson: Heavy ion ranges at 100 keV in aluminum. Phys. Letters *46A*, 45-46 (1973).	R. 100 keV Cs, Ba, La, Sm, Eu, Tb, Au $\rightarrow$ Al

708	I. Kh. Lemberg, A. A. Pasternak: New method of investigating the electronic and ion-atomic mechanisms of stopping heavy ions in matter. Zh. ETF Pis. Red. *19*, 784-87 (1974). [Engl. trans. JETP Letters *19*, 401-02 (1974).]	S. 18 MeV Cd → Cd, 12.5 MeV Ni → Ni
709	H. L. Ottosen: Specialeopgave Aarhus University, p. 1-54 (1974) (in Danish).	S. 0.6-2.5 MeV p → Al
710	J. J. Grob, A. Grob, A. Pape, P. Siffert: Energy loss of heavy ions in nuclear collisions in silicon. Phys. Rev. B*11*, 3274-79 (1975).	S, δS, $\eta(\varepsilon)$. 0.3-2 MeV N, Si → Si (cryst.)
711	V. N. Avdeichikov, G. F. Gridnev, O. V. Lozhkin, N. A. Perfilov: The limiting energy resolution of thin dE/dx detectors for 5-9 MeV alpha particles. Bull. Acad. Sci. USSR. Phys. Ser. *34*, No. 1, 190-96 (1970).	δS. 5-8.7 MeV α → Si
712	V. G. Telkovskii, V. I. Pistunovich: Passage of ions of various gases through a thin silver film. Dokl. Akad. Nank. SSSR *113*, 1035-38 (1957). (Sov. Phys. Doklady *2*, 184-86 (1957).)	S. 2-20 keV H, He, C, N, O → Ag
713	G. Dearnaley, G. A. Gard, W. Temple, M. A. Williams: Depth distribution of gallium ions implanted into silicon crystals. Appl. Phys. Letters *27*, 17-18 (1975).	R, δR. 40 keV ^{72}Ga → Si (cryst.)
714	R. Behrisch, J. Bøttiger, W. Eckstein, U. Littmark, J. Roth, B. M. U. Schertzer: Implantation profiles of low-energy helium in niobium and the blistering phenomena. Appl. Phys. Letters *27*, 199-201 (1975).	R, δR. 15 keV ^{3}He → Nb
715	H. H. Andersen, J. Bøttiger, H. Wolder Jørgensen: Ranges of ions with $Z_1 \geq 54$ in Al and Al$_2$O$_3$. Appl. Phys. Letters *26*, 678-79 (1975).	R, δR. 75 keV ^{129}Xe, 80 keV ^{153}Eu, ^{197}Au, 100 keV ^{205}Tl → Al, Al$_2$O$_3$ 75 keV ^{133}Cs → Al
716	M. Iwaki, K. Gamo, K. Masuda, S. Namba, S. Ishihara, I. Kimura, K. Yokota: Experimental method for measuring both atom and carrier concentration profiles in the same sample of ion-implanted silicon layers by radioactive-ion implantation. Nucl. Instr. Methods *127*, 93-98 (1975).	R, δR. 45 keV As → Si
717	P. E. Thompson, R. B. Murray: Ion bombardment of alkali halides. I. Range and damage profiles of protons in KCl. Rad. Effects *25*, 127-32 (1975).	R. 0.5-15 MeV p → KCl
718	A. Barcz, A. Turos, L. Wielunski, W. Rosinski, B. Wojtowicz-Natanson: Depth distribution of silver ions implanted in Si and SiO$_2$. Rad. Effects *25*, 91-96 (1975).	R, δR. 20-140 keV ^{107}Ag → Si, SiO$_2$
719	G. W. Carriveau, G. Beauchemin, E. J. Knystautas, E. H. Pinnington, R. Drouin: Energy loss measurements of low energy ions in thin carbon foils. Phys. Letters *46A*, 29-30 (1973).	S. rel. to 60 keV p. 100, 200 keV N, Ne, Ar, Mn, Kr, Xe → C
720	T. Andersen, O. H. Madsen, G. Sørensen: Beam-foil spectroscopy at low initial ion energies. Physica Scripta *6*, 125-30 (1972).	S. 90-300 keV Xe → C

721	V. M. Pistryak, A. K. Gnap, V. F. Kozlov, R. I. Garber, A. I. Fedorenko, Ya. M. Fogel': Concentration profile of boron ions implanted into silicon with energies of 30 and 100 keV. Fiz. Tverd. Tela *12*, 1281-82 (1970). [Engl. trans. Sov. Phys. Solid State *12*, 1005-06 (1970).]	R, δR. 30, 100 keV B $\rightarrow$ Si (cryst.)
722	P. E. Schambra, A. M. Rauth, L. C. Northcliffe: Energy loss measurements for heavy ions in mylar and polythene. Phys. Rev. *120*, 1758-61 (1960).	S. 12-120 MeV C, 16-160 MeV O, 20-200 MeV Ne $\rightarrow$ mylar, polythene
723	R. A. Schmitt, R. A. Sharp: Measurement of the range of recoil atoms. Phys. Rev. Letters *1*, 445-47 (1958).	R. 130 keV ^{11}C $\rightarrow$ polystyvene, 85 keV ^{18}F $\rightarrow$ teflon, 45 keV ^{34}Cl $\rightarrow$ saran, 33 keV ^{45}Ti $\rightarrow$ Ti; 30 keV ^{53}Fe $\rightarrow$ Fe, 25 keV ^{63}Zn $\rightarrow$ Zn, 25 keV ^{62}Cu, ^{64}Cu $\rightarrow$ Cu; 16 keV ^{91}Mo $\rightarrow$ Mo, 14 keV ^{106}Ag $\rightarrow$ Ag, 9 keV ^{196}Au $\rightarrow$ Au
724	P. Zahn: Energieverlust von α-Rückstosskernen in Formvar. Z. Physik *175*, 85-98 (1963).	S. 169 keV ^{208}Pb $\rightarrow$ Formvar
725	H. H. Andersen: *Studies of atomic collisions in solids by means of calorimetric techniques*. Aarhus University. Aarhus p. 1-279 (1974).	S. 5-17 MeV p, d $\rightarrow$ Al, Cu
726	C. M. Lattes, P. H. Fowler, P. Cuer: A study of the nuclear transmutations of light elements by the photographic method. Proc. Phys. Soc. *59*, 883-900 (1947).	R. 1.2-13.1 MeV p, 2.1-13.0 MeV α $\rightarrow$ Emulsion
727	W. D. Warters, W. A. Fowler, C. C. Lauritsen: The elastic scattering of protons by lithium. Phys. Rev. *91*, 917-21 (1953).	S. rel. to 952 keV p.200-1300 keV p $\rightarrow$ Li
728	J. Rotblatt: Range-energy relation for protons and alpha-particles in photographic emulsion for nuclear research. Nature *167*, 550-51 (1951).	R. 0.2-16.4 MeV p, 1.1-18.9 MeV α $\rightarrow$ Emulsion
729	B. R. Nielsen: Specialeopgave Aarhus University, p. 1-75 (1975) (in Danish).	S. 1.6-6.3 MeV p, 1.6-9 MeV d, $\rightarrow$ Al, Ag; 2.2-18.6 MeV ^{3}He $\rightarrow$ Al, 3.0-13.5 MeV ^{3}He $\rightarrow$ Ag, 3.2-17.2 MeV α $\rightarrow$ Al, 3.2-19.2 MeV α $\rightarrow$ Ag
730	H. D. Maccabee, M. R. Raju, C. A. Tobias: Fluctuations of energy loss in semiconductor detectors. IEEE Trans. Nucl. Sci. NS*13*, No. 6, 176-79 (1966).	δS. 730 MeV p, 910 MeV α $\rightarrow$ Si
731	J. Català, W. M. Gibson: Range-energy relation for protons and alpha-particles in photographic emulsions for nuclear research: Nature *167*, 551-52 (1951).	R. 5-16.3 MeV p, 8-19 MeV α $\rightarrow$ Emulsion
732	J. Rotblatt: Range-energy relation for protons and alpha particles in photographic emulsions for nuclear research. Nature *165*, 387-88 (1950).	R. 1-8 MeV p, α $\rightarrow$ Emulsion

733	H. Bradner, F. M. Smith, W. H. Barkas, A. S. Bishop: Range-energy relation for protons in nuclear emulsion. Phys. Rev. *77*, 462-67 (1950).	R. 17-39.5 MeV p → Emulsion
734	C. M. G. Lattes, P. H. Fowler, P. Cuer: Range-energy relation for protons and alpha-particles in the new Ilford "Nuclear Research" emulsion. Nature *159*, 301-02 (1947).	R. rel. to air. 2-13 MeV p, 5-9 MeV α → Emulsion
735	W. K. H. Panofsky, F. L. Fillmore: The scattering of protons by protons near 30 MeV, photographic method. Phys. Rev. *79*, 57-70 (1950).	R. 10.8-12.0 MeV p → Emulsion
736	N. Nereson, F. Reines: Nuclear emulsions and the measurement of low energy neutron spectra. Rev. Sci. Instr. *21*, 534-545 (1950).	R. 0.2-1.5 MeV p → emulsion
737	E. M. Gunnersen, G. James: On the efficiency of the reaction $H^3(d,n)He^4$ in titanium tritide bombarded with deuterons. Nucl. Instr. Methods *8*, 173-84 (1960).	S. 40-120 keV p → TiH
738	H. Faraggi: Mesure Précise de l'Énergie des Particules Lourdes Chargées des Faibles Parcours par Impregnation d'Émulsion Photographiques. Ann. Physique *6*, 325-400 (1951).	R. 0.7-3.2 MeV p, 1-5.3 MeV α, 0.85-1.8 MeV ^{7}Li, 40 keV C → Emulsion
739	A. H. Armstrong, G. M. Frye, Jr.: The Reaction $B^{11}(n,\alpha)Li^8(\beta^-)Be^{8*}(2\alpha)$ for 12- to 20-MeV neutrons. Phys. Rev. *103*, 335-40 (1956).	R. 0.5-7 MeV ^{8}Li → Emulsion
740	R. A. Peck, Jr.: A calibration for Eastman proton plates. Phys. Rev. *72*, 1121 (1947).	S. 2.5-9 MeV p → Emulsion
741	K. Locher, P. Stoll: The (γ,α) reaction of B^{11} and B^{12}. Phys. Rev. *90*, 164-65 (1953).	R. 1.6-5.3 MeV Li → Emulsion
742	H. T. Richards, V. R. Johnson, F. Ajzenberg, M. J. W. Laubenstein: Proton range-energy relation for Eastman NTA emulsions. Phys. Rev. *83*, 994-95 (1951).	R. 1-17 MeV p → Emulsion
743	J. C. C. Tsai, J. M. Morabito: In-depth profile detection limits of nitrogen in GaP and nitrogen, oxygen, anf fluorine in Si by SIMS and AES, in S. Namba (edt.): *Ion Implantation in Semiconductors*, Plenum, N.Y. p. 115-24 (1975).	R, δR. 50 keV N → GaP, Si; 50 keV O, F → Si
744	S. Ludvik, L. Scharpen, H. E. Weawer: Measurement of arsenic implantation profiles in silicon using electron spectroscopic technique. Idem. 155-62 (1975).	R, δR. 25, 40 keV ^{75}As → Si
745	M. Iwaki, K. Gamo, K. Masuda, S. Namba, S. Ishihara, I. Kimura, K. Yokota: Atom and carrier profiles in As implanted Si. Idem. 163-68 (1975).	R, δR. 25 keV As → Si
746	Y. Ohmura, K. Koike, H. Kobayashi, K. Murakami: Deviated gaussian profiles of implanted boron and deep levels in silicon. Idem. 183-88 (1975).	R, δR. 60-200 keV ^{10}B, ^{11}B → Si

747	K. Wittmaack, F. Schulz, B. Hietel: Range distributions of boron in silicon dioxide and the underlying substrate. Idem. 193-200 (1975).	R, δR. 5-150 keV ^{11}B → SiO_2
748	J. Biersack, D. Fink: Implantation of boron and lithium in semiconductors and metals. Idem. 211-18 (1975).	R, δR. 50-150 keV ^{10}B → Si
749	P. F. Engel, F. Chernow: Deep penetration of implanted Po in CdS. Idem. 267-74 (1975).	R, δR. 25 keV ^{210}Po → CdS (cryst.)
750	T. Hirao, T. Ohzone, S. Takayanagi, H. Horumi: Ion implantation in polycrystalline silicon. Idem. 599-604 (1975).	R, δR. 50-150 keV B → Si
751	P. Blank, K. Wittmaack: Range parameter distortion in heavy ion implantation. Phys. Letters *54A*, 33-34 (1975).	R, δR. 20-50 keV Xe → Si
752	A. Nomura, S. Kiyono: Stopping power of copper, silver and gold for protons and helium ions of low energy. J. Phys. D: Appl. Phys. *8*, 1551-59 (1975).	S. 4-16 keV H, He → Cu, Ag, Au
753	K. Gamo, M. Iwaki, K. Masuda, S. Ishihara, K. Kimura, S. Namba: Concentration profiles of ion implanted impurities into silicon. Sci. Pap. Instr. Phys. Chem. Res. (Japan) *65*, 19-21 (1971).	R, δR. 20-80 keV In, 20 keV Sb → Si
754	J. Rickards: Energy straggling of protons in carbon. Nucl. Instr. Methods *127*, 397 (1975).	δS. 460 keV p → C
755	T. R. Ophel, G. W. Kerr: A study of the energy loss of 0.36 to 4.5 MeV protons in thin carbon films. Nucl. Instr. Methods *128*, 149-55 (1975).	S, δS. 0.36-4.5 MeV p → C
756	J. Bøttiger, S. T. Picraux, N. Rud: Depth profiling of hydrogen and helium isotopes in solids by nuclear reaction analysis; in O. Meyer, G. Linker, F. Käppeler (edt): *Ion Beam Surface Layer Analysis*. Plenum, New York, p. 811-19 (1976).	R. 12 keV H → Al_2O_3
757	J. D. Melvin, T. A. Tombrello: Energy loss of low energy protons channeling in silicon crystals. Rad. Effects *26*, 113-26 (1975).	S. 0.5-1.6 MeV p → Si (cryst.)
758	A. Antilla, M. Bister, J. Keinonen: DSA lifetimes in ^{21}Na and ^{23}Na derived from experimental stopping parameters. Z. Physik A *274*, 227-32 (1975).	R. 20-100 keV ^{23}Na → Ta
759	R. Kelly: Sputtering and depth-distribution phenomena in KCl, Al_2O_3, TiO_2. Can. J. Phys. *46*, 473-85 (1968).	R. 10 keV Kr → KCl, TiO_2, Al_2O_3
760	J. Almodovar, E. Paez-Mozo, R. Gaeta: Range in tungsten of Ba-140 fragments from the thermal neutron-induced fission of enriched uranium. J. Inorg. Nucl. Chem. *29*, 2831-37 (1967).	R. 100 MeV ^{140}Ba → W
761	N. Bloembergen, P. J. van Herden: The range and straggling of protons between 35 and 120 MeV. Phys. Rev. *83*, 561-66 (1951).	R, δR. 35-120 MeV p → Al, Cu, Pb

762	C. J. Cook, E. Jones Jr., T. Jorgensen, Jr.: Range-energy relations of 10- to 250-keV protons and helium ions in various gases. Phys. Rev. *91*, 1417-22 (1953).	R. 4-250 keV p → H_2, 7-250 keV p → Ar, Air, N_2, 8-250 keV p → CO, 10-250 keV p → CH_4, 13-250 keV p → O_2, 20-250 keV α → H_2, Ar, Air
763	W. M. Gibson, D. J. Prowse, J. Rotblat: Range-energy relation in nuclear track emulsions for protons of energy up to 21 MeV. Nature *173*, 1180-81 (1954).	R. 2-21 MeV p → Emulsion
764	E. S. Borovik, N. P. Katrich, G. T. Nikolaev: The determination of the penetration coefficient of fast H^+ ions in metals by the condensation method. Atomnaya Energiya *23*, 102-05 (1967). [Engl. trans. Soviet Atom. Energ. *23*, 793-96 (1967).]	R. 35 keV p → Ni, Ti
765	V. N. Lepishinskaya, E. M. Zarutskiy: Penetration of medium energies ions into metals. *Proceeding of the 8th Int. Conf. Ion. Phenom. in Gases*, Springer. Wien, p. 44 (1967).	R, S. 1-16 keV H^+, H_2^+, H_3^+ → Cu
766	T. Jokic: Ranges of 20-80 keV xenon ions in copper crystals in G. Musa, I. Ghica, A. Popescu and L. Nastase (eds.) *Phenomena in Ionized Gases 1969*, Editura Academie Bucharest, p. 96 (1969).	R. 20-100 keV Xe → Cu (cryst.)
767	V. V. Makarov, N. N. Petrov: Penetration of 2 to 11 keV lithium ions into silicon carbide single crystals. Izv. Akad. Nauk. SSSR Ser. Fiz. *30*, 890-91 (1966). [Engl. trans. Bull. Acad. Sci. USSR, Phys. Ser. *30*, 925-26 (1966).]	R. 2-11 keV Li → SiC (cryst.)
768	N. I. Venikov, N. I, Chumakov: The time of flight method for measuring the range-energy relation for He ions in aluminum at 18 to 38 MeV. Atomnaya Energiya *17*, 503-04 (1964). [Engl. trans. Sov. Atom. Energ. *17*, 1269-70 (1964).]	R. 18-38 MeV α → Al
769	P. V. Pavlov, E. I. Zorin, D. I. Tetelbaum, Ya. S. Popov: The depth of penetration and the distribution of radiation defects in germanium bombarded by argon and nitrogen ions. Fiz. Tverd. Tela *6*, 3222-26 (1964). [Engl. trans. Sov. Phys. Solid State *6*, 2577-80 (1964).]	R. 46-82 keV Ar, N → Ge
770	J. Mösner, G. Schmidt, J. Schintlmeister: Energie-Reichweite-Kurve für ^{7}Li-Ionen in Luft. Ann. Physik *18*, 268-70 (1966).	R. 0.1-5.0 MeV Li → Air
771	L. Wieninger: Über die Reichweiten von Polonium α-Strahlen in einigen Alkalihalogenid-Kristallen, (NaCl, KCl, KBr, KJ). Acta Physica Austriaca *4*, 355-59 (1950).	R. 5.3 MeV α → NaCl, KCl. KBr, KJ (all cryst.)
772	N. P. Katrich: Effect of ion energy and depth distribution of interstitial hydrogen in a nickel film on the hydrogen ion range. Atomnaya Energiya *26*, 286-87 (1969). [Engl. trans. Sov. Atom. Energy. *26*, 318-19 (1969).]	R. 14-42 keV p → Ni

773	A. Garin, H. Faraggi: Parcours des Alpha de 4.5 MeV dans L'Uranium, L'Or, Le Zirconium et le Silicium. J. Phys. Radium *19*, 76-78 (1958).	R. 4.5 MeV $\alpha \rightarrow$ Si, Zr, Au, U
774		
775	T. D. Lagerlund, M. Blecher, K. Gotow, D. Jenkins, W. C. Lan: Measurement of range and range straggling of low energy pions and muons. Nucl. Instr. Methods *128*, 525-29 (1975).	R, δR. 39.3 MeV μ^+, 33.0, 37.6 MeV π^+, $\pi^- \rightarrow$ Al
776	Yu. M. Kushnir, A. N. Kabanov, L. N. Krumiakova: Measurement of energy loss in gases of 70 keV lithium ions by an electrostatic analyser. Radioteknika i Elektronika *5*, (1960). [Engl. trans. Rad. Eng. Elect. Phys. *5*, 197-____ (1960).]	S. 70 keV Li $\rightarrow$ He, Ar, N_2, O_2
777	S. Datz, T. S. Noggle, C. D. Moak: Anisotropic energy losses in a face-centered cubic crystal for high-energy ^{79}Br and ^{127}I ions. Phys. Rev. Letters *15*, 254-57 (1965).	S. 12-70 MeV ^{79}Br, 23-82 MeV ^{127}I $\rightarrow$ Au (cryst.)
778	W. E. Miller, J. A. Hutchby: Stopping cross sections for 0.25-3.0 MeV ^{4}He ions in cadmium sulfide. J. Appl. Phys. *46*, 4479-83 (1975).	S. 0.25-3.0 MeV $\alpha \rightarrow$ CdS
779	H. H. Andersen, J. F. Bak, H. Knudsen, B. R. Nielsen: Stopping Powers of Al, Cu, Ag, and Au for MeV Hydrogen, helium, and Lithium ions. Z_1^3, and Z_1^4 proportional deviations from the Bethe formula. Preprint Aorhus (1977).	S. 1.6-6.8M p, 1.6-9M d, 2.2-18.6M ^{3}He, 3.2-17.2M α, 14-20M ^{6}Li, 8-23M ^{7}Li $\rightarrow$ Al, 1.6-7.2M p, 2.3-8.5M d, 5-18M α, 10-17M ^{7}Li $\rightarrow$ Cu, 1.6-6.3M p, 1.6-9M d, 3.0-12.5M ^{3}He, 3.2-19.2M α, 8.5-21M ^{7}Li $\rightarrow$ Ag, 1.6-5M p, 1.7-8.4M d, 4.1-15.5M α, 13-15M ^{6}Li, 10-21M ^{7}Li $\rightarrow$ Au.
780	A. Grob, J. J. Grob, P.Siffert: Energy loss and straggling of heavy ions by nuclear interactions in silicon. Nucl. Instr. Methods *132*, 273-79 (1976).	S, δS, $\nu(\varepsilon)$. 300-2000 keV ^{12}C, ^{14}N, ^{16}O, ^{20}Ne, ^{28}Si, ^{32}S, ^{40}Ar $\rightarrow$ Si
781	R. Ishiwari, N. Shiomi, S. Shirai: Z_1^3 effect on the stopping powers of several metallic elements for 28.8 MeV alpha particles: Deviations of experimental data from theories. Phys. Letters *51A*, 54-54 (1975).	S. 28.8 MeV $\alpha \rightarrow$ Al, Ti, Fe, Ni, Cu, Mo, Ag, Ta, Au
782	F. G. Neskev, A. A. Puzanov, K. S. Shyshkin, E. I.Sirotinin, A. F. Tulinov, G. D. Kedmanov: The determination of energy losses of nitrogen ions from backscattering spectra. Rad. Effects *25*, 271-73 (1975).	S. 1.0-7.4 MeV N $\rightarrow$ Ti, Ge, Ni, Ag, Au, W
783	R. Ishiwari, N. Shiomi, T. Katayama-Kinoshita, F. Sawada-Yasue: Search for possible geometrical effect on stopping power measurement. J. Phys. Soc. Jap. *39*, 557-65 (1975).	S. 8.78 MeV $\alpha \rightarrow$ Al, Cu, Ag, Ta
784	R. A. Langley, R. S. Blewer: Measurement of the stopping cross sections for protons and ^{4}He ions in erbium and erbium oxide: A test of Bragg's rule. Nucl. Instr. Methods *132*, 109-16 (1976).	S. 0.25-2.5 MeV p, $\alpha \rightarrow$ Er, Er_2O_3

785	R. A. Langley: Stopping cross sections for helium and hydrogen in H_2, N_2, O_2 and H_2S (0.3-2.5 MeV). Phys. Rev. B *12*, 3575-83 (1975).	S. 0.3-2.5 MeV p, $\alpha \rightarrow H_2$, N_2, O_2, H_2S
786	B. M. U. Schertzer, P. Borgesen, M.-A. Nicolet, J. W. Mayer: Determination of stopping cross sections by Rutherford backscattering; in O. Meyer, G. Linker, F. Käppeler (edt) *Ion Beam Surface Layer Analysis*. Plenum, N.Y., p. 33-46 (1976)	S. 0.2-2.0 MeV $\alpha \rightarrow$ Au, Pt, Ta_2O_5, SiO_2
787	S. N. Chumilov, N. Ya. Rutkevitch and A. P. Klioutcharev: The range-energy relation for carbon ions in NiKFi nuclear emulsion (in French). p. 242-44 in P. Demers (edt.) *Photographie Corpusculaire*, Vol. 3 Presses Universitaires de Montreal (1964).	R. 40-112 MeV C $\rightarrow$ Emulsion
788	I. Kh.Lemberg, V. I. Medvedev and A. V. Plavko: The range-energy relation for NiKFi Ya 2 emulsion (in French). Idem. p. 245-48 (1964).	R. 3-16 MeV $\alpha \rightarrow$ Emulsion
789	R. Buhl, W. K. Huber and E. Löbach: Messung von Konzentrationsprofilen dünner Schichten mit der Sekundärionen-massenspektrometrie (SIMS).Vakuum-Technik *24*, 189-94 (1975).	R. 60, 100 keV B $\rightarrow$ Si
790	M. Brendle, F. Gugel, G. Steidle: The ranges of alpha particles in H_2, He, CH_4 and CO_2 at energies from 0.5 to 5.3 MeV. Nucl. Instr. Methods *130*, 253-256 (1975).	R. 0.5-5.3 MeV $\alpha \rightarrow H_2$, He, CH_4, CO_2
791	R. Gähler, J. Kalus, R. Behrisch: A measurement of the first moment of the range distribution of (n, 2n) recoils in Au and Nb. Nucl. Instr. Methods *130*, 203-06 (1975).	R. 71 keV ^{196}Au $\rightarrow$ Au, 143 keV ^{92}Nb $\rightarrow$ Nb
792	B. Efken, D. Hahn, D. Hilscher, and G. Wüstefeld: Energy loss and energy loss straggling of N, Ne and Ar ions in thin targets. Nucl. Instr. Methods *129*, 219-225 (1975).	S, δS. 10 MeV N $\rightarrow N_2$; 15 MeV Ne $\rightarrow$ He, N_2, SF_6, Ar; 10, 15 MeV Ne $\rightarrow$ C; 5-15 MeV Ar $\rightarrow N_2$, C; 15 MeV Ar $\rightarrow SF_6$, Ar
793	G. D. Kerr, L. M. Hain, N. Underwood and A. W. Walthers: Molecular stopping cross sections of air, N_2, Kr, CO_2 and CH_4 for alpha particles. Health Physics *12*, 1475-80 (1966).	S. 0.3-5 MeV $\alpha \rightarrow N_2$, Air, Kr, CO_2, CH_4
794	L. P. Moroz and A. Kh. Ayukhanov: Comparative study of the depth of penetration of different ions into dielectric films by the method of secondary ion-electron emission. Izv. Akad. Nauk SSSR Ser. Fiz. *28*, (1964). [Engl. trans: Bull Acad. Sci. USSR Phys. Ser. *28*, 1301-05 (1964).]	R. 0.43-1.58 keV Na, 1.58-3.04 keV Rb $\rightarrow$ RbCl
795	A. Hurrle and G. Sixt: Cesium profiles in silicon and in SiO_2-Si double layers as determined by SIMS measurements. Applied Physics *8*, 293-302 (1975).	R. 10-100 keV Cs $\rightarrow$ Si

172

796	M. Kiselevich, A. Lyatushinski, V. Zhuk and B. P. Osipenko: Implantation of heavy ions into silicon single crystals. Fiz. Tverd. Tela *17*, 1080-84 (1975). [Engl. trans. Sov. Phys. Solid State *17*, 687-89 (1975).]	R, δR. 45 keV ^{82}Sr, ^{128}Ba, ^{134}Ce, ^{140}Nd, ^{145}Eu, ^{149}Gd, ^{152}Tb, ^{160}Er, ^{167}Tm, ^{169}Lu $\rightarrow$ Si (cryst. chann. and random)
797	J. Bøttiger, J. R. Leslie and N. Rud: Range profiles of 6-16-keV hydrogen ions implanted in metal oxides. J. Appl. Phys. *47*, 1672-75 (1976).	R, δR. 6-16 keV H $\rightarrow$ Al_2O_3, Nb_2O_5, Ta_2O_5
798	D. G. Simons, D. J. Land, J. G. Brennan and M. D. Brown: Range, distribution and stopping power of 800-keV ^{14}N$^+$ ions implanted in metals from $Z_2 = 22$ to $Z_2 = 32$. Phys. Rev. A*12*, 2383-92 (1975).	R, δR, S. 800 keV $\rightarrow$ Ti, V, Cr, Mn, Fe, Co, Ni, Cu, Zn, Ga, Ge
799	Yu. V. Bulgakov, V. S. Nikolaev and V. I. Shulga: Impact-parameter dependence of inelastic energy losses for He and N ions channeled in Si. Phys. Stat. Sol. (a) *31*, 341-50 (1975).	S, δS. 1.3, 6.6 MeV α, 4.4 MeV N $\rightarrow$ Si (cryst.)
800	D. A. Leich and T. A. Tombrello: A technique for measuring hydrogen concentration versus depth in solid samples. Nucl. Instr. Methods *108*, 67-71 (1973).	F, δR. 11.5 keV H $\rightarrow$ SiO_2 (cryst. and amorph.), Feld spar
801	E. Ligeon A. Guivarc'h: A new utilization of ^{11}B ion beams: hydrogen analysis by ^{1}H(^{11}B,α)$\alpha\alpha$ nuclear reaction. Rad. Effects. *22*, 101-105 (1974).	R, δR. 10 keV H $\rightarrow$ Si
802	W. K. Hofker: Implantation of boron in silicon. Philips Res. Repts. No. 8, p. 1-121 (1975).	R,δR. 30-800 keV B $\rightarrow$ Si (amorph., polycryst.)
803	P. D. Scholten, M. R. Skokan, K. W. Kemper, W. G. Moulton: Range and distribution of Gd ions implanted in Nb thin films. Phys. Rev. B*13*, 42-44 (1976).	R,δR. 50-150 keV Gd $\rightarrow$ Nb
804	T. Sharma, S. Mukherji: Ranges of ^{140}Ba from thermal neutron induced fission of ^{235}U along the different crystallographic directions of aluminum single crystals. J. Inorg. and Nucl. Chem. *37*, 1845-49 (1975).	R. 68.5, 140 MeV ^{140}Ba $\rightarrow$ Al (cryst., polycryst.)
805	J. Roth, R. Behrisch, B. M. U. Schertzer: Determination of the depth distribution of implanted helium atoms in niobium by Rutherford backscattering. Appl. Phys. Letters *25*, 643-44 (1974).	R,δR. 4 keV He $\rightarrow$ Nb (cryst.; chann. and random)
806	M. T. Guseva, A. N. Mansurova: Radiation-enhanced diffusion of boron in germanium during ion implantation. Rad. Effects *20*, 207-10 (1973).	R,δR. 30 keV B $\rightarrow$ Si
807	H. B. Dietrich, W. H. Weisenberger, J. Comas: Anomalous migration of ion-implanted Al in Si. Appl. Phys. Letters *28*, 182-84 (1976).	R,δR. 60 keV Al $\rightarrow$ Si

808	D. J. Land, D. G. Simons, J. G. Brennan, M. D. Brown: Unfolding techniques for the determination of distribution profiles from resonance reaction gramma-ray yields. in O. Meyer, G. Linker, F. Käppeler (edt): *Ion Beam Surface Layer Analysis*, Plenum, New York, p. 851-61 (1976).	R,δR. 800 keV N $\rightarrow$ Z_2 = 22-32, 40-42
809	V. S. Vavilov, V. V. Galkin, V. V. Krasnopevtsev, Yu. V. Milyutin: Distribution of the conductivity with depth in diamond doped by bombardment with 10-45 keV Li[7] ions. Fiz. Tekh. Poluprovodnikov, *4*, 1180-82 (1970). [Engl. trans. Sov. Phys. Semicond. *4*, 1000-02 (1970)]	R. 10-45 keV Li $\rightarrow$ Diamond
810	V. S. Vavilov, M. A. Gukasyan, E. A. Konorova, Yu. V. Milyutin: Implantation of antimony ions into diamond. Fiz. i Tekh. Poluprovodnikov *6*, 2384-91 (1972). [Engl. trans. Sov. Phys. Semicond. *6*, 2384-91 (1973)].	R. 40 keV Sb $\rightarrow$ C (diamond)
811	B. Blanchard, J. L. Combasson, J. C. Bourgoin: Boron-implanted profiles in diamond. Appl. Phys. Letters, *28*, 7-8 (1976).	R,δR. 40-250 keV B $\rightarrow$ C (diamond)
812	U. Hauser, W. Neuwirth, W. Pietsch, K. Richter: The "Inverted Doppler Shift Attenuation" method: A new method for the investigation of atomic and molecular charge distributions. Radiochem. Radioanal. Letters, *18*, 301-09 (1974).	S. 175 keV Li $\rightarrow$ CrB, CrB_2, NbB, NbB_2, MoB, MoB_2, TaB, TaB_2, WB, W_2B_5
813	W. Neuwirth, W. Pietsch, K. Richter, U. Hauser: Electronic stopping cross sections of elements and compounds for swift Li ions.. Z. Physik A *275*, 209-14 (1975).	S. 80-840 keV Li $\rightarrow$ Be, B, Al, Ti, Cu, Ta, AlB_2, AlB_{12}, B_4C, B_2O_3, BPO_4, B_4Si, CaB_6, CeB_6, CrB, CrB_2, Cr_2B_3, H_2O, D_2O, HBO_2, H_3BO_3, HfB_2, KBF_4, KBH_4, LaB_6, $LiBH_4$, MoB, MoB_2, $Na_2B_4O_7$, $Na_2B_4O_7$.10 H_2O (borax), $NaBH_4$, NbB, NbB_2, NH_4BF_4, TaB, TaB_2, VB_2, WB, W_2B_5, YB_6, ZrB_2
814	U. Hauser, W. Neuwirth, W. Pietsch, K. Richter: On the determination of collision cross sections by nuclear Doppler shift. Z. Physik *269*, 181-88 (1974).	S. 80-840 keV Li $\rightarrow$ CrB_2
815	W. Pietsch, U. Hauser, W. Neuwirth: Stopping powers from the inverted Doppler shift attenuation method: Z-oscillations, Bragg's rule or chemical effects; solid and liquid state effects. Nucl. Instr. Methods *132*, 79-87 (1976).	S. 70, 100 keV Li $\rightarrow$ B; 100 keV Li $\rightarrow$ Al, Ti, Cu, Ta, C, TaB, TaB_2, CrB, CrB_2, NbB, NbB_2, MoB, MoB_2, TaB, TaB_2, WB, W_2B_5, 300 keV C $\rightarrow$ Nb, Mo, Ta, Ag; 100 keV Li $\rightarrow$ LiOH, HCl, KNO_3, H_2SO_4, KF, KCl, KBr, KJ, LiCl, $LiNO_2$, $ZnCl_2$, $Zn(NO_3)_2$, $LaCl_3$, $La(NO_3)_2$, $CeCl_3$, $Ce(NO_3)_2$. (All as electrolytes, dep. on conc.)
816	J. S.-Y. Feng: Energy loss of 0.3-2.0 MeV [4]He ions in aluminum. J. Appl. Phys. *46*, 444-45 (1975).	S. 0.3-2.0 MeV α $\rightarrow$ Al.

817	E. Ligeon, A. Guivarc'h: Hydrogen implantation in silicon between 1.5 and 60 keV. Rad. Effects *27*, 129-37 (1976).	R, δR. 1.5-60 keV H → Si. S. 1.5-60 keV H, 2.0 MeV ^{11}B → Si
818	M. B. Al-Bedri, S. J. Harris, H. G. S. F. Parish: Energy loss and straggling measurements for low energy protons transmitted through thin solid films. Rad. Effects *27*, 183-87 (1976)	δS. 0.6-1.6 MeV p → Melinex, Al, Cu, Au, Pb
819	J. P. S. Pringle: A comparison of sectioning methods used to measure concentration profiles in anodic oxides. Can. J. Phys. *54*, 56-65 (1976).	R. δR. 10-80 keV ^{24}Na 20-160 keV ^{85}Kr, ^{125}Xe → Al_2O_3, Ta_2O_5, WO_3. 20-160 keV ^{41}Ar → Al_2O_3, 20-80 keV ^{42}K → Ta_2O_5, 20-80 keV ^{41}Ar → WO_3
820	W. K. Lin, S. Matteson, D. Powers: α-particle stopping cross section of gold and silver as measured from thick targets. Phys. Rev. B *10*, 3746-55 (1974).	S. 0.3-2.0 MeV α → Au, Ag
821	J. S. Forster, D. Ward, H. R. Andrews, G. C. Ball, G. J. Costa, W. G. Davies, I. V. Mitchell: Stopping power measurements for ^{19}F, ^{24}Mg. ^{27}Al, ^{32}S and ^{35}Cl at energies 0.2 to 3.5 MeV/nucleon in Ti, Fe, Ni, Cu, Ag and Au. Nucl. Instr. Methods *136*, 349-59 (1976).	S. 2.2 MeV p, 0.2-3.5 MeV/amu ^{19}F, ^{24}Mg, ^{27}Al, ^{32}S, ^{35}Cl → Ti, Fe, Ni, Cu, Ag, Au
822	G. Thieme: Bestimmung des elektronischen Energieverlustes von H$^+$-, He$^+$- und N$^+$-ionen in Gold durch Vergleich von Messergebnissen mit Monte-Carlo-Rechnungen. Vakuum-Technik *25*, 5-12 (1976).	S. 40-110 keV H, He, N → Au
823	J. S. -Y. Feng, W. K. Chu, M. -A. Nicolet: Stopping-cross-section additivity for 1-2-MeV ^{4}He$^+$ in solid oxides. Phys. Rev. B *10*, 3781-88 (1974).	S. 1-2 MeV α → MgO, Al_2O_3, SiO_2, α-Fe_2O_3, Fe_3O_4
824	W. A. Lanford, H. P. Trautvetter, J. F. Ziegler, J. Keller: A new precision technique for measuring the concentration versus depth of hydrogen in solids. Appl. Phys. Letters *28*, 566-68 (1976).	R, δR. 7.5 keV p → Si (cryst.)
825	H. Ryssel, H. Müller, K. Schmid: Damage dependent electrical activation of boron implanted silicon. Applied Physics *3*, 321-24 (1974).	R, δR. 34 keV B → Si
826	D. K. Sood, G. Dearnaley: Ion-implanted surface alloys in copper and aluminum. in G. Carter, J. S. Colligon, W. A. Grant (edts.): *Applications of Ion Beams to Materials*. Institute of Physics Conf. Series No. 28 p. 169-203 (1976).	R. 150 keV Au, Mo, Bi; 200 keV Ta, Cu; 250 keV Rb; 300 keV Ru, Cs, Ce, Eu → Cu. 150 keV Mo, Gd, Bi; 200 keV Ag, Cu, Se, Au; 250 keV Rb; 300 keV Cd, Cs → Al.
827	H. Mitsui: Measurement of the distribution of fission fragments recoiling into aluminum. J. Nucl. Sci. Technol. *2*, 481-84 (1965).	R. 97 MeV ^{95}Zr, ^{95}Nb; 140 MeV ^{140}Ba, ^{140}La → Al

828	E. Wenger, R. P. Gardner, K. Verghese: Molecular stopping cross sections of alpha particles in butane, propane, ethane, neon, helium, and hydrogen. Health Physics 25, 67-71 (1973).	S. 2.54-4.93 MeV $\alpha \rightarrow H_2$, 2.09-5.29 MeV $\alpha \rightarrow Ne$, 0.9-5.13 MeV $\alpha \rightarrow He$, 0.5-5.11 MeV α C_2H_6, 0.59-4.44 MeV $\alpha \rightarrow C_3H_8$, 0.82-4.22 MeV $\alpha \rightarrow C_4H_{10}$
829	E. G. Wikner, H. Horiye, D. K. Nichols: Elastic versus inelastic energy loss of recoil germanium and silicon atoms. Phys. Rev. 136, A1428-32 (1964).	S, $\eta(\varepsilon)$. 68-157 keV Si $\rightarrow$ Si, 19-36 keV Ge $\rightarrow$ Ge
830	J. Kalus: Abbremsung von Chrom-Atomen mit einer Energie von 20-90 eV in verschiedenen Substanzen. Z. Naturforsch. 20a, 391-94 (1965).	R. 20-90 eV Cr $\rightarrow$ V, V_2O_5, VOC_2O_4, $V(CH(COCH_3)_2)_3$ $V(CH(COCH_3)_2)_2$
831	R. Bader, S. Kalbitzer: Low energy boron and phosphorus implants in silicon. (B). Doping profiles. Rad. Effects 6, 211-16 (1970).	R, δR. 6 keV B, 15 keV P $\rightarrow$ Si (cryst.).
832	J. Bohdansky, J. Roth, W. P. Poschenrieder: The trapping of hydrogen ions in zirconium for ion energies between 0.3 and 6 keV, in G. Carter, J. S. Colligon, W. Grant (edts.): *Applications of Ion Beams to Materials*. Institute of Physics Conf. Series No. 28. London. p. 307-12 (1976).	R, δR. 1, 4 keV D $\rightarrow$ Zr
833	H. Albrecht, H. Munzel: Spezifischer Energie-verlust von schweren Ionen in Aluminum, Silber und Gold. Z. Physik 220, 381-91 (1969).	S. Fission fragments $\rightarrow$ Al, Ag, Au
834	J. Mory: Parcours Moyen des Fragments de Fission dans Quelques Métaux avec le Mica Comme Détecteur. Rev. Phys. Appl. 3, 387-95 (1968).	S. Fission fragments $\rightarrow$ Al, Ti, Fe, Ni, Cu, Mo, Ag, Au
835	H. Okabayashi, D. Shinoda: Lateral spread of ^{31}P and [11]B ions implanted in silicon. J. Appl. Phys. 44, 4220-21 (1973).	R, δR, $\delta R_{\perp}$. 145, 260 keV ^{31}P; 80, 150 keV [11]B $\rightarrow$ Si
836	Y. Akasaka, K. Horie, S. Kawazu: Lateral spread of boron ions implanted in silicon. Appl. Phys. Letters 21, 128-29 (1972).	$\delta R_{\perp}$. 75-250 keV B $\rightarrow$ Si
837	A. Feuerstein, S. Kalbitzer, H. Oetzmann: Range parameters of heavy ions at 10 and 35 keV in silicon. Phys. Letters 51A, 165-66 (1975).	R, δR. 10-35 keV Ge, As $\rightarrow$ Si
838	M. Bertin, M. Bruno, G. Vannini, A. Vitale: Measurement of the energy loss in argon of fission fragments having selected masses. Phys. Letters 43A, 231-33 (1973)	S. Fission fragments $\rightarrow$ Ar
839	J. E. Greene, F. Sequeda-Osorio, B. G. Streetman, J. R. Noonan, G. G. Kirpatrich: Measurement of boron impurity profiles in Si using glow discharge optical spectroscopy. Appl. Phys. Letters 25, 435-38 (1974).	R, δR. 120 keV B $\rightarrow$ Si
840	C. Chasman, K. W. Jones, H. W. Kramer, W. Brandt: Band-gap effects in the stopping of Ge[72*] atoms in germanium. Phys. Rev. Letters 21, 1430-33 (1968).	S. $\eta(\varepsilon)$. 10-30 keV Ge $\rightarrow$ Ge

841	Y. Ohmura, K. Koike: Evidence for electronic stopping in ion implantation: Shallower profile of lighter isotope ^{10}B in Si. Appl Phys. Letters *26*, 221-22 (1975).	R, δR. 50-200 keV ^{10}B, ^{11}B $\rightarrow$ Si
842	J. K. Hirvonen, G. K. Hubler: Application of a high-resolution magnetic spectrometer to near-surface analysis, in O. Mayer, G. Linker and F. Käppeler (edts.): *Ion Beam Surface Layer Analysis*, Plenum, N. Y. p. 457-69 (1976).	R, δR. 2-60 keV ^{209}Bi, 60 keV ^{69}Ga $\rightarrow$ Si
843	W. A. Grant, J. S. Williams, D. Dodds: Measurement of projected and lateral range parameters for low energy heavy ions in silicon by Rutherford backscattering. Idem, p. 235-44 (1974).	R, δR, δR$_\perp$. 10-80 keV Pb, 50-400 keV Bi, 40 keV Ar, Cu, Kr, Cd, Al, Dy, W $\rightarrow$ Si
844	A. Feuerstein, G. Grahmann, S. Kalbitzer, H. Oetzmann: Rutherford backscattering analysis with very high depth resolution using an electrostatic analysing system. Idem p. 471-81 (1976).	δS. 100-200 keV p,d; 250 keV α $\rightarrow$ Pt, Au, SiO$_2$
845	H. Oetzmann, A. Feuerstein, H. Grahmann, S. Kalbitzer: Range parameters of heavy ions in silicon and germanium with released energies from $0.001 < \varepsilon < 10$. Idem p. 245-54 (1976).	R, δR. 1-40 keV Bi $\rightarrow$ Si, 5-60 keV Sb, As $\rightarrow$ Si; 60 keV Au $\rightarrow$ Si, Al; 40 keV Ge $\rightarrow$ Si; 1 keV Bi $\rightarrow$ Ge
846	A. l'Hoir, C. Cohen, G. Amsel: Experimental study of the stopping power and energy straggling of MeV ^{4}He, ^{12}C, ^{14}N and ^{16}O ions in amorphous aluminum oxide. Idem. p. 965-76 (1976).	S, δS. 0.3-1.7 MeV α, ^{12}C, ^{14}N, ^{16}O $\rightarrow$ Al$_2$O$_3$
847	M. Hufschmidt, W. Möller, V. Heintze, D. Kamke: Depth profiling of deutererons in metals at large implantation depths using the nuclear reaction technique. Idem p. 831-40 (1976).	R, δR. 100-400 keV d $\rightarrow$ Ni
848	D. G. Simons, D. J. Land, J. G. Brennan, M. D. Brown: Z_2 dependence of the electronic stopping power of 800 keV ^{14}N+ ions in targets from carbon through molybdenum. Idem p. 863-71 (1976).	S. 800 keV N $\rightarrow$ Z_2 = 22-32, 40-42
849	W. Eckstein, R. Behrisch, J. Roth: Achieved depth resolution in profiling light atoms by nuclear reactions. Idem p. 821-30 (1976).	R, δR. 15 keV ^{3}He $\rightarrow$ Nb (cryst. chann.) 1.5 keV ^{3}He $\rightarrow$ Nb (cryst. random)
850	J. Roth, R. Behrisch, W. Eckstein, B. M. U. Schertzer: Depth profiling of implanted ^{3}He in solids by nuclear reaction and Rutherford backscattering. Idem p. 47-54 (1976).	R, δR. 15 keV ^{3}He $\rightarrow$ Nb (cryst.) S. 500 keV d $\rightarrow$ Nb
851	J. F. Ziegler, W. K. Chu, J. S. -Y. Feng: Evidence of solid state effects in the energy loss of ^{4}He ions in matter. Idem p. 15-27 (1976).	S. 2 MeV α $\rightarrow$ Fe$_2$O$_3$, Fe$_3$O$_4$, MgO, Al$_2$O$_3$, SiO$_2$, Si$_3$N$_4$
852	J. S. Williams: The application of low angle Rutherford backscattering to surface layer analysis. Idem p. 223-34 (1976)	R, δR. 5 keV Cr $\rightarrow$ Ge, 20 keV Pb $\rightarrow$ Si
853	D. Olmos, F. Aldape, J. Cavillo, A. Chi, S. Romero, J. Rickards: Energy dependence of proton straggling in carbon. Idem p. 65-74 (1976).	δS. 0.46-4.79 MeV p $\rightarrow$ C

854	M. Luomajärvi, A. Fontell, M. Bister: Energy straggling of ^{4}He ions in Al and Cu in the backscattering geometry. Idem. p. 75-85 (1976).	S, δS. 0.5-2 MeV $\alpha \to$ Al, Cu
855	B. H. Armitage, P. N. Trehan: Energy loss straggling of protons in thick absorbers. Idem p. 55-63 (1976) [identical to 866]	δS. 5-12 MeV p $\to$ Al, V, Ni, Mo, Ag, Ta, Au
856	R. S. Blewer: Depth distribution and migration of implanted helium in metal foils using proton backscattering, in S. T. Picraux, E. P. EerNisse, F. L. Vook (edts.): *Applications of Ion Beams to Metals*. Plenum, N.Y. p. 559-72 (1974)	R, δR. 54-158 keV He $\to$ Cu
857	R. S. Blewer: Depth distribution of implanted helium and other low-z elements in metal films using proton backscattering. Appl. Phys. Letters *23*, 593-95 (1973).	R, δR. 54-158 keV He $\to$ Cu
858	S. Hasegawa, H. Ishiwara, S. Furukawa, T. Shimizu: The lattice location of phosphorus atoms implanted into silicon. Jap. J. Appl. Phys. *15*, 391-92 (1976).	R. 100 keV P $\to$ Si
859	J. Bogancs, S. Deme, J. Gyulai, A. Nagy, V. M. Nazarov, A. Csoke, Yu. S. Yazvitsky: A method for the determination of boron ranges in ion implanted silicon by the ^{10}B(n, α) nuclear reaction. Joint Institute for Nuclear Research. Dubna, Report No. P. 14-8295 (1974) (in Russian).	R. 20-80 keV B $\to$ Si
860	K. C. Shane, H. Lanner, G. G. Seaman: Energy loss of low-energy ^{40}Ca ions in carbon. J. Appl. Phys. *47*, 2286-88 (1976).	S. 121-335 keV ^{40}Ca $\to$ C
861	H. B. Dietrich, L. E. Plew: ^{19}F range-energy curve in Si from 100 keV - 550 keV. J. Appl. Phys. (1976)	R. 100-550 keV F $\to$ Si
862	R. B. Strittmatter, B. W. Wehring: Alpha-particle energy straggling in solids. Nucl. Instr. Methods *135*, 173-77 (1976).	δS. 6.11 MeV $\alpha \to$ Al, Ag, Au
863	P. Armbruster, K. Sistemich, J. P. Bocquet, Ch. Chauvin, Y. Glaize: Energy straggling of heavy ions (A $\approx$ 100, E/A $\approx$ 1 MeV/amu) in solids and gases, the limiting factor of the charge resolving power of energy-loss detectors. Nucl. Instr. Methods *132*, 120-32 (1976).	δS. 90 MeV Kr, Rb $\to$ Si; 84 MeV Y, Zr $\to$ C; 83 MeV Kr $\to$ Ar
864	G. Varley, J. C. Willmott, F. Kearns: The energy losses of heavy ions. Nucl. Instr. Methods *135*, 167-72 (1976).	S. 70-250 MeV Nd $\to$ Nd, Dy $\to$ Dy, Er $\to$ Er, 180-300 MeV Kr $\to$ Nd, Dy, Er
865	G. E. Hoffmann and D. Powers: Energy straggling of α particles in solid materials. Phys. Rev. A *13*, 2042-48 (1976).	δS. 0.5-2.0 MeV $\alpha \to$ Ti, Cr, Co, Cu, Ag
866	B. H. Armitage, P. N. Trehan: Energy loss straggling of protons in thick absorbers. Nucl. Instr. Methods *134*, 359-62 (1976) [identical to 855]	δS. 6-12 MeV p $\to$ Al, V, Ni, Mo, Ag, Ta, Au

178

867	K. -H. Schmidt, H. Wohlfarth, H. -G. Clerc, W. Lang, H. Schrader, K. E. Pferdekämper: Energy loss, energy straggling and angular straggling of heavy ions in carbon foils. Nucl. Instr. Methods *134*, 157-66 (1976).	S, δS. 80-100 MeV Kr, Rb, Sr, Y, Zr, Nb, Sb, Te → C
868	W. W. M. Allison, C. B. Brooks, J. N. Bunch, R. W. Flemming, R. K. Yamamoto: The ionisation loss of relativistic charged particles in thin gas samples and its use for particle identification. II. Experimental Results. Nucl. Instr. Methods *133*, 325-34 (1976).	S, δS. 25-150 GeV/c p, π → Ar + 20% CO_2
869	J. W. Tape, W. M. Gibson, J. Remillieux, R. Laubert, H. E. Wegner: Energy loss of atomic and molecular ion beams in thin foils. Nucl. Instr. Methods *132*, 75-77 (1976).	S. 0.3-1.0 MeV/atom H^+, H_2^+; 1.6-2.9 MeV/atom O^-, O_2^- → C
870	C. D. Moak, B. R. Appleton, J. A. Biggerstaff, M. D. Brown, S. Datz, T. S. Noggle, H. Verbeek: The velocity dependence of the stopping power of channeled iodine ions from 0.6 to 60 MeV. Nucl. Instr. Methods *132*, 95-98 (1976).	S. 0.6-60 MeV I, 0.2-5.0 MeV Kr → Ag (cryst.)
871	G. Grahmann, S. Kalbitzer: Nuclear and electronic stopping powers of low energy ions with Z ≤ 10 in silicon. Nucl. Instr. Methods *132*, 119-23 (1976).	S. 2-60 keV H, He, B, C, N, Ne → Si
872	K. G. Prasad, R. P. Sharma: Energy loss of channeled protons in Al single crystals. Nucl. Instr. Methods *132*, 103-07 (1976).	S.∿1.5 MeV p → Al (cryst.)
873	R. A. Baragiola, D. Chivers, D. Dodds, W. A. Grant, J. S. Williams: Ranges in silicon of ions with atomic numbers 62 ≤ Z_1 ≤ 66 at 100 keV. Phys. Letters *56A*, 371=73 (1976).	R, δR. 100 keV ^{152}Sm, ^{153}Eu. ^{157}Gd, ^{159}Tb, ^{164}Dy → Si
874	O. Fich, J. A. Golovchenko, K. O. Nielsen, E. Uggerhoj, C. Vraast-Thomsen, G. Charpak, S. Majewski, F. Sauli, J. Ponpon: Ionization loss of channeled 1.36 GeV/c protons and pions. Phys. Rev. Letters, *36*, 1245-48 (1976).	S. 1.35 GeV/c p, π → Ge (cryst. chann. and random)
875	G. Mende, G. Küster: Die Abtragung dünner Schichten von P- und B-implantierten Silizium mit Hilfe der anodischen Oxydation. Thin Solid Films *35*, 215-20 (1976).	R, δR. 30 keV P → Si
876	V. Martini: Stopping cross section measurements with heavy ions in the keV range in gases by time-of-flight spectroscopy. Preprint. Nucl. Instr. Methods (1976).	S. 20-156 keV Pb → H_2
877	N. N. Demidovich, I. E. Nakhutin, V. G. Shatunov: Certain features of the stopping power of gases for fission fragments. Pis'ma Zh. Eksp. Teor. Fiz. *22*, 526-28 (1975) [Eng. Trans. JETP Letters *22*, 257-58 (1975)].	S. Fission fragments → Air
878	A. A. Bednyakov, Yu. V. Bulgakov, V. S. Nikolaev, V. P. Sobakin, B. M. Popov: Stopping power distribution for fast helium and nitrogen ions passing through metal films. Zh. Eksp. Teor. Fiz. *68*, 2067-74 (1975) [Engl. trans. Sov. Phys. JETP *41*, 1034-37 (1976)]	δS. 1.3 MeV α, 4.6 MeV N → Al, Cu, Ag, Au

879	S. Matteson, E. K. L. Chau, D. Powers: Stopping cross section of bulk graphite for α particles. Phys. Rev. A *14*, 169-75 (1976).	S. 0.3-2.0 MeV $\alpha \rightarrow$ C
880	J. F. Ziegler, W. K. Chu, J. S. -Y Feng: Empirical corrections to the energy loss of ^{4}He ions in oxides. Appl. Phys. Letters *27*, 387-90 (1975).	S. 2 MeV $\alpha \rightarrow Fe_2O_3$, Fe_3O_4, MgO, Al_2O_3, SiO_2, Si_3N_4 all rel. to metal
881	D. C. Santry, C. W. Sitter: Range and retention studies of 40-keV ions in solids, in H. Wagner, W. Walcher (edts.) *Proc. Int. Conf. Elmagn. Isotope Separators and their Techniques*. Marburg, p. 505-24 (1970).	R, δR. 40 keV ^{14}C, ^{16}O, ^{32}P, ^{58}Co, ^{204}Tl, $\rightarrow$ Au, ^{24}Na, ^{32}P, ^{58}Co, ^{65}Zn, ^{75}Se, ^{85}Kr, ^{181}Hf, ^{134}Cs, ^{110m}Ag, ^{131}I, ^{133}Xe $\rightarrow$ W; ^{125}Xe $\rightarrow WO_3$
882	V. V. Avdeichikov, E. A. Ganza, O. V. Lozhkin: Energy resolution of thin semiconductor ΔE detectors for alpha particles and heavy ions. Nucl. Instr. Methods *131*, 61-68 (1976).	δS. 0.9-7.7 MeV α, 8-40 MeV ^{12}C, 5-48MeV ^{14}N, 2-100 MeV ^{15}N, 9-20 MeV ^{16}O, 8-30 MeV ^{20}Ne, 11-60 MeV ^{22}Ne, 12-15 MeV ^{40}Ar $\rightarrow$ Si
883	U. Bill, R. Sizmann, C. Varelas, K. E. Rehn: Transition of axial to planar channeling. Rad. Effects *27*, 59-66 (1975).	S, δS. 100 MeV S $\rightarrow$ Si (cryst.)
884	F. Centmayer, R. Sizmann: Thermal formation of the super-tail in the implantation profile of krypton in single crystalline tungsten. Rad. Effects *28*, 49-55 (1976).	R, δR. 27 keV Kr $\rightarrow$ W (cryst.)
885	H. H. Andersen, P. Hornshøj, L. Hojshølt-Poulsen, H. Knudsen, B. R. Nielsen, R. Stensgaard: A simple energy-calibration procedure for electrostatic accelerators. Nucl. Instr. Methods *136*, 119-24 (1976).	S. 1.5-6.5 MeV p, 2.6-8.5 MeV d $\rightarrow$ Ag
886	M. Braun, B. Emmoth, R. Buchta: Concentration profiles and sputtering yields measured by optical radiation of sputtered particles. Rad. Effects *28*, 77-83 (1976).	R, δR. 50-120 keV Al $\rightarrow$ Ag; 60 keV Na $\rightarrow$ Si
887	W. A. Grant, J. S. Williams, D. Dodds: Measurement of the lateral spread of heavy ions implanted into silicon. Rad. Effects *29*, 189-90 (1976).	$\delta R_\perp$. 35 keV Cu; 40 keV Cd, Xe, Dy; 45 keV Kr, W; 10-40 keV Pb; 45-400 keV Bi $\rightarrow$ Si
888	D. Fuller, J. S. Colligon, J. S. Williams: The application of correlated SIMS and RBS techniques to the measurement of ion implanted range profiles. Surf. Science *54*, 647-58 (1976).	R. δR. 20, 30 keV ^{133}Cs $\rightarrow$ Si; 20 keV ^{133}Cs $\rightarrow$ Al
889	A. S. Rudnev, V. I. Rolyakov, E. I. Sirotinen, A. F. Tulinov: A study of relative energy losses of channeled protons in a tungsten single crystal. Phys. Stat. Sol. (a) *35*, K23-27 (1976).	S, δS. 6.3 MeV p $\rightarrow$ W (cryst.)
890	H. Matsumura, S. Furukawa: Simple formulation for energy straggling of helium in silicon. Rad. Effects *27*, 245-46 (1976).	δS. 1-2 MeV $\alpha \rightarrow$ Si

891	J. S. Colligon, D. Fuller: Secondary ion emission studies of the range profiles of implanted ions. Rad. Effects *28*, 183-187 (1976).	R, δR. 20 keV ^{133}Cs $\rightarrow$ Cu, Al
892	J. Mönkedick: Channeling-und Blocking Effekte bei Transmission von Deuteronen der Energie 16 bis 27.5 MeV durch Lithiumfluorid-Einkristalle. Z. Physik A *278*, 5-8 (1976).	S, δS. 16-27.5 MeV d $\rightarrow$ LiF (cryst.)
893	M. E. Helmbold: Energieverlust und Lichtausbeute bei axialem Channeling von α-Teilchen in Anthrazen. Z. Physik A *278*, 9-14 (1976).	S, δS. 6.05, 8.78 MeV α $\rightarrow$ Anthracene
894	H. H. Andersen, J. F. Bak, H. Knudsen, P. Moller-Petersen, B. R. Nielsen: Experimental investigations of higher-order Z_1 corrections to the Bethe stopping power formula, in B. Navinsek (edt.) *Physics of Ionized Gases 1976*. Contributed Papers. J. Stefan Institute. Ljubljana. p. 221-23 (1976).	S. 3-6.8 MeV d, 5-13 MeV α, 8.5-21 MeV ^{7}Li $\rightarrow$ Ag, Au
895	K. Güttner, S. Hofman, D. Marx, G. Münzenberg, F. Nickel: Range and range straggling of heavy ions in solids. Idem p. 228-29 (1976).	R, δR 37 MeV Pr $\rightarrow$ Ni, Ta; 49, 91 MeV Pd $\rightarrow$ Nd
896	J. C. Eckhardt: Energy loss and straggling of protons and helium ions traversing some thin solid foils. Phys. Rev. (1976).	S, δS. 20-260 keV H, He $\rightarrow$ Ge, Se, Pd, Ag, Sb, Bi
897	P. V. Ramana Murthy, G. D. Demeester: The use of gas proportional counters to distinguish protons from pions in the cosmic radiation of energies near or greater than 100 GeV. Nucl. Instrum. Methods *56*, 93-105 (1976).	δS. 1.5 GeV/c μ, 80 MeV p, 1.5, 40 GeV/c π^- $\rightarrow$ Ar+7% CH_4
898	P. V. Ramana Murthy: Relativistic rise of the most probable energy loss in a gas proporitonal counter. Nucl. Instrum. Methods *63*, 77-82 (1968).	S. 5-12 GeV/c π^- $\rightarrow$ Ar + 7% CO_2
899	A. S. Lodhi, D. Powers: Energy loss of α-particles in gaseous C-H and C-H-F compounds. Phys. Rev. A *10*, 2131-40 (1974).	S. 0.3-2.0 MeV α $\rightarrow$ $C_2H_2F_2$, $C_2H_4F_2$, C_3H_8, C_4H_6, C_4H_{10}, C_3H_4
900	M. T. Shehata, R. Kelly: The formation and structure of anodic films on beryllium. J. Electrochem. Soc. *122*, 1359-65 (1975).	R. δR, 30 keV $\rightarrow$ Be
901	Han-chuan Liu Cheng: Range energy measurements in gases. Thesis. University of Southern California (1965).	R. 2-8 MeV α $\rightarrow$ N_2, O_2, Air, Ar
902	J. H. Thorngate: Measurements of energy losses, distribution of energy loss and additivity of energy losses for 50 to 150 keV protons in hydrogen and nine hydrocarbon gases. ORNL/TM-5165 (1976).	S, δS. 50-150 keV H $\rightarrow$ H_2, CH_4. CHCH, CH_2CH_2, CH_3CH_3, CH_3CCH, CH_2CCH_2, CH_3CHCH_2, $CH_2CH_2CH_2$, $CH_3CH_2CH_3$
903	B. J. Farmer, H. Bichsel: Range energy measurements for 2- to 5-MeV protons in Ni and Ag. Bull. Am. Phys. Soc. *5*, 263 (1960).	R. 2-5.2-MeV p $\rightarrow$ Ni, Ag
904	C. Tschalär, H. Bichsel: Mean excitation potential of light compounds. Phys. Rev. *175*, 476-8 (1968).	R. 3-30 MeV p $\rightarrow$ Si, Al, SiO_2, Al_2O_3, $C_3H_5O_2$

905	I. M. Vasilievskii, I. I. Karpov, V. I. Petrushkin, Yu. D. Prokoshkin: Proton ranges amd ionization energy losses in various materials. Yaderna Fizika *9*, 997-1008 (1968) [Eng. transl. Sov. J. Nucl. Phys. *9*, 583-9 (1969)].	R. 660 MeV p → C, Al, Cu, Sn, Pb
906	N. H. Harley, B. S. Pasternack: Experimental absorption applied to lung dose from thoron daughters. Health Physics *24*, 379-86 (1976).	S. 0.2-8.5 MeV α → Polycarbonate
907	N. H. Harley, B. S. Pasternack: Alpha absorption measurements applied to lung dose from radon daughters. Health Physics *23*, 771-82 (1972).	S. 0.5-7.5 MeV α → Polycarbonate
908	H. H. Andersen, J. F. Bak, H. Knudsen, P. Møller-Petersen, B. R. Nielsen: Experimental investigation of higher-order Z_1 corrections to the Bethe stopping-power formula. Nucl. Instrum. Methods (1977).	S. 1.6-4.6 MeV p, 3.4-8.4 MeV d, 6-12 MeV α, 13-15 MeV ^{6}Li. 10-21 MeV ^{7}Li → Au, 2.8-6.5 MeV d, 5-13 MeV α, 8.5-21 MeV ^{7}Li → Ag, 3-5.8 MeV d, 5-11 MeV α, 10-17 MeV ^{7}Li → Cu, 2.4-6.8 MeV d, 4.4-13.6 MeV α, 14-20 MeV ^{6}Li, 8-23 MeV ^{7}Li → Al
909	C. D. Moak: Experiments with heavy ions. Nucl. Instrum. Methods *28*, 155-9 (1964).	S. 22-115 MeV I → C, Al, Ni, Au
910	A. Nomura, S. Kiyono: Measurements of energy distribution of low energy light ions through copper film and its statistical analysis. Jap. J. Appl. Phys. *15*, 1773-7 (1976).	S. 5-10 keV H → Cu, δS.2-13 keV H, 6-12 keV He → Cu
911	H. K. Abele, P. Glässel, P. Mair-Komor, H. J. Scheerer, H. Rösler, H. Vonach: A method for measuring the uniformity of thin targets by means of an alpha source and a Q3D spectrograph. Nucl. Instrum. Methods *137*, 157-67 (1976).	δS. 8.78 MeV α → C, Au, Ni, Al, SiO$_2$, Ru
912	M. J. Geary, A. K. M. M. Haque: The stopping power and straggling for alpha particles in tissue equivalent materials. Nucl. Instrum. Methods *137*, 151-5 (1976).	S, δS. 0.5-5.5 MeV α → (73.7% H$_2$, 25% CO$_2$, 1.3% N$_2$), (63.4% CH$_4$, 33.4% CO$_2$, 3% N$_2$), Air, Melinex, Mylar
913	J. H. Thorngate: Measurements of distributions of energy loss for 51, 102, and 153 keV protons in nine hydrocarbon gases. Nucl. Instrum. Methods *137*, 569-75 (1976).	δS. 50-150 keV H → CH$_4$, CHCH, CH$_2$CH$_2$, CH$_3$CH$_3$, CH$_3$CCH$_2$, CH$_3$CHCH$_3$, CH$_2$CH$_2$CH$_2$, CH$_3$CH$_2$CH$_3$
914	H. Kräutle: Study of the sputtering process with Rutherford backscattering. Nucl. Instrum. Methods *137*, 553-7 (1976).	R, δR, 50 keV Au → Al, 5-30 keV B → Si
915	G. Frick, C. Gehringer, B. Heusch, Ch. Ricaud, P. Wagner, E. Baron: Stripping study for the "GANIL" project. IEEE Trans. Nucl. Sci. NS*23*, 1137-9 (1976).	δS. 100 MeV Au, 110 MeV I, 48 MeV Ni → C

916	F. Harris, T. Katsura, S. Parker, V. Z. Peterson, R. W. Ellsworth, G. B. Yodh, W. W. M. Allison, C. B. Brooks, J. H. Cobb, J. H. Mulvey: The experimental identification of individual particles by the observation of transition radiation in the x-ray region. Nucl. Instrum. Methods 107, 413-22 (1973).	δS. 3-GeV/c π^- → Ar + 7% Co_2
917	V. Heintze: Die Richtungsabhängigheit von Deuteronenverteilungen in mono- und polykrystallinen Nickel-Absorbern. Z. Physik B 25, 159-165 (1976).	R. δR, $\delta R_\perp$. 400 keV d → Ni (cryst. and polycryst.)
918	G. P. Pokhil, A. S. Rudnev, E. I. Sirotinen, A. F. Tulinov: Energy losses of protons moving in the planar channel. Rad. Effects 30, 167-70 (1976).	S. 6.3 MeV p → W (cryst. chann. to random ratio)
919	P. T. Callaghan, P. Kittel, N. J. Stone, P. D. Johnson: Impurity-site distribution of implanted Bi in iron and nickel studied by channeling and nuclear orientation. Phys. Rev. B 14, 3722-31 (1976).	R, δR, 200 keV Bi → Fe, Ni (cryst. chann. and random)

TABLE VI

Author Index

A

Abele, H. K. 911
Abroyan, I. A. 473
Addamiano, A. 499
Aditya, P, K. 43
Adlerholz, M. 625, 626
Aitken, D. W. 385
Ajzenberg, F. 742
Akasaka, Y. 541, 692, 836
Akimura, H. 291, 399
Al-Bedri, M. B. 472, 818
Albrecht, H. 833
Aldcroft, D. A. 316
Aldape, F. 853
Alexander, D. 360
Alexander, G. 49
Alexander, J. M. 185, 204, 557
Allen, G. 9
Allison, S. K. 2, 177, 176, 370
Allison, W. W. M. 868, 916
Almodovar, J. 760
Altman, M. 257
Ambrox, R. 669
Ambrosi, D. A. 400
Amme, R. C. 238
Amos, T. M. 372
Amsel, G. 846
Anashkina, É. S. 590
Andersen, H. H. 205, 269, 280, 358, 374, 404, 499, 715, 725, 779, 885, 893, 908
Andersen, T. 338, 415, 460, 720
Anderson, G. S. 297
Anderson, M. G. 600
d'Andlam, C. A. 100
Andreen, C. J. 225, 284, 290, 343
Andreev, V. N. 405, 409
Andrews, H. R. 821
Anianson, G. 3, 560
Anton, D. 370
Anttila, A. 461, 490, 758
Apel, P. 655
Appleton, B. R. 21, 224, 257, 305, 384, 483, 572, 573, 576, 595, 631, 673, 870
Appleyard, R. K. 4
Arakawa, E. T. 206
Aras, N. K. 699
Arkhipov, E. P. 410
Armbruster, P. 863
Armbruster, R. 279, 387

184

Armitage, B. H. 322, 855, 866
Armstrong, A. H. 739
Aron, W. A. 14
Arora, M. R. 678
Artemov, A. D. 701
Atterling, A. 75
Augenlicht-Jakobson, D. 662
Avdeichikov, V. V. 628, 711, 882
Axtmann, R. C. 259
Ayukhanov, A. Kh. 794

B

Bach, H. 417, 610
Badanoin, M. 459
Bader, M. 7, 8
Bader, R. 831
Badinka, C. 1
Baglin, J. E. E. 453
Baily, N. A. 392, 401, 425, 667
Bak, J. F. 779, 894, 908
Bakker, C. J. 218
Baldinger, E. 245
Ball, G. C. 169, 173, 202, 613, 821
Ball, M. P. 649
Baragiola, R. A. 873
Barcz, A. 718
Barile, S. 9
Barkan, S. 167, 580
Barkas, W. H. 207, 214, 221, 585, 733
Barker, P. H. 235, 688
Barnaby, C. F. 41
Barnett, C. F. 48, 57
Baron, E. 301, 915
Barrett, J. H. 483, 576
Barrett, P. H. 585
Baskin, R. 52
Bason, F. 390
Bates, L. F. 10
Batholomew, G. A. 33
Bätzner, H. 407
Baylis, W. E. 433
Beachemin, G. 636, 719
Becker, J. 115
Bednyakov, A. A. 67, 878
Beetzhold, W. 536
Behrisch, R. 508, 576, 714, 791, 805, 849, 850
Bellamy, E. 321
Bellamy, J. C. 95
Bellavance, G. R. 681
Bentini, G. G. 132, 463, 480, 574
Benton, E. V. 231
Bergdolt, A. M. 279
Bergström, I. 154, 298
Bernard, J. 537
Bernhard, F. 395
Bernstein, W. 658
Berthold, F. 210
Bertin, M. 838

C

Furukawa, S. 125, 510, 540, 858, 890
Furuya, T. 690

G

Gabriele, S. A. 623
Gaeta, R. 760
Gähler, R. 609, 791
Gaissmair, B. 661
Galaktionov, V. V. 514
Galkin, V. V. 809
Galloni, R. 705
Gamo, K. 527, 544, 716, 745, 753
Ganza, E. A. 628, 882
Garber, R. 721
Garbincius, P. H. 445
Garcia-Munoz, M. 177
Gard, G. A. 27, 334, 378, 413
Gardes, D. 685
Gardner, R. P. 828
Garfinkel, A. F. 269
Garin, A. 773
Geary, M. J. 912
Gehringer, C. 915
Geiger, J. S. 434
Geiger, K. W. 251
Geller, K. N. 15
Gen, M. Ya. 561
Gérardin, G. 565
Gerthsen, C. 53
Gettings, M. 501
Ghosh, S. K. 73, 74
Gibbons, J. F. 243, 333
Gibson, W. M. 731, 762
Gibson, W. M. 189, 224, 257, 305, 343, 532, 533, 595, 869
Gilat, J. 185, 204
Gilber, F. C. 549
Gilbert, C. W. 54
Gittings, R. P. 616
Giusti, P. 623
Glaize, Y. 863
Glässel, P. 911
Gnap, A. K. 721
Gobeli, G. W. 56
Golanski, A. 702
Golovchenko, J. A. 874
Goode, P. D. 27
Gooding, T. J. 535
Goodman, P. 119
Gorodetzky, S. 279, 387
Gotow, K. 635, 775
Gott, Yu. V. 159, 410, 651
Götz, G. 577
Govind, P. K. 697
van de Graff, R. J. 306
Graham, R. L. 160, 434
Grahmann, H. 844, 845, 871
Grant, I. S. 195
Grant, W. A. 611, 843, 873

H

M

N

Q

R

W